THE THAMES AND H

GENERAL EDIT

Television graphics

8

Ron Hurrell

The Thames and Hudson Manual of Television graphics

with 255 illustrations, in colour and black and white

Thames and Hudson T&H London

The author's thanks are due to Mr Dave Thomas for the pictures on pp. 58, 61, 62 and 63; to Mr Andrew Lee for the picture on p. 59 (top); to Mr Terry Gilliam for the pictures on pp. 34 and 110; and to the Mansell Collection for the picture on p. 53 (below).

Printed in Great Britain by Jarrold and Sons Ltd, Norwich

ISBN 0 500 67004 8 Clothbound
0 500 68004 3 Paperback

Contents

1 Introductory

The problem of designing visual symbols for the television screen points logically to the development of a new and exciting form of imagery. This particular branch of design is unique; it is a form of television and of graphic art, containing aspects of each. It extends the range of both, and adds to them its own breadth of expression. Not only is there a need to transmit clear and unmistakable information at any point in the day's programming, but often it is a matter of expressing ideas in a way the live image cannot achieve. Captions and titles are a basic requirement of television; there is a necessity to describe exactly what follows and what has come before. There must be a way of reporting quickly and exactly the weather, sports results, news and service announcements between programmes, of subtly suggesting the passage of time or changing situations.

The written language is the stock-in-trade of television graphic design – adapted, stylized and applied to varying situations as it is in other fields of graphic design – qualified by the emphasis of tone and pattern, and by pictorial symbolism where the meaning can be extended, made more exact or awaken understanding more quickly or completely by such means. Added to this are the factors of time and movement implicit in the design of graphics. While it is possible to illustrate the developing frames of an animated title or the successive sequences of a roller caption with full understanding, it is almost impossible to impart the motivation or force of a title fading or cutting into the programme. This can be appreciated only in its real context. The subtleties of meaning are developed through the accompanying sounds or speech, and cannot be distinguished by the eye alone. Also it is not possible to show the experience or thought-process of the designer who produced the graphics. The nearest means to this is to outline the ideal experience or background of a television designer.

The basic prerequisites are creative ability (which can be brought to some form of fruition and awareness through art college training) and enough humility to start from the bottom rung, in order to learn the practical and technical requirements of television design. This framework of rules, limitations and techniques, which must be closely observed, gives the designer a conscious and subconscious starting-point, the segment of vision and design within which there is endless scope for fresh thought and experiment. Allied to this is the need for experience of work in commercial design, illustration, typography and layout, lettering, photography, model-making and art direction. A quality which is also very important, but which is very hard to define in concrete terms, is the ability to be in sympathy with the design subjects, the varying types of programmes and the medium used to illustrate them.

Commercial art experience is necessary in order to produce, with speed, graphics which are bold and simple, which state the necessary information without any visual verbosity, and with a clear tonal contrast. (Because of the structure of a television image, built up as it is from a series of horizontal lines, there is always a loss of definition in any design.) This experience enables the designer to take short cuts, use tricks which he knows will be applicable to television and save valuable time. Speed and accuracy are not just assets but the necessities of the designer. Much of the work involves illustration, the condensation of ideas, significant points, even the total atmosphere of a programme, into one or two designs. This problem makes full demands on the designer's heightened powers of perception, which must be coupled with a strong sense of design and the ability to portray and synthesize abstract forms into concrete, recognizable terms. This rests entirely on the designer's ability as a draughtsman.

Knowledge of typography and layout, especially in relation to printing processes, is useful when one considers the tremendous output of lettered captions. Even the simplest programme needs titles and credit captions, which must display good layout and that balance of weight which is necessary for legibility. The standard demanded of a typographer in a printing house is the basis of all lettering produced for television. This needs to be coupled with good taste and a genuine feeling for the style and content of the programme. The first images can make or mar the viewing potential of any programme, therefore the quality of the typography must be perfect: second-best will not do.

Although much of the lettering used in television is of a mechanical form – hot-press and instant lettering – there is a demand for hand-drawn lettering. It is important for the designer to know the principles and practice of creative and characterful high-quality lettering, so that specialized titling styles may be produced which are suitable for a particular series of programmes or subjects. There is always room for development in hand lettering, so long as the shortcomings and pitfalls of existing styles are avoided, and so long, of course, as this form of lettering is not allowed to consume too much valuable time. Hand lettering is a skill which comes with practice, but it is preferable for the designer to have acquired some proficiency before he starts in television. The ideal is for hand lettering to be as fluent and almost as quick as the designer's own handwriting.

Photography is another skill the designer should learn. Although in most departments of television there is a separate photography unit, a knowledge of the photographic processes is very useful. Many programme titles, whether they are still or animated sequences, demand the combination of photography and lettering. Obviously if the designer has the ability to choose and shoot his own photographs, he can have complete control of the finished design. Even if he cannot take his own photographs, the designer must know how to give explicit instructions to the photographic department and not to make impossible demands.

The designer will find any experience (even from childhood or school) of modelling with clay or similar substances, the construction of balsa-wood aircraft, scenery for model railways and so on, invaluable in model-making. There is a demand for models, working constructions and experiments in television. This may be catered for by a special department, but if the designer can produce them himself, it can save a lot of time. Not only are they used for their own sake within programmes, but they can be the basis for photographs and titles. There are nowadays many materials at the designer's disposal for the speedy making of models with lightness and strength, able to simulate reality. Apart from the construction of three-dimensional models, there is the making of special studio animations. This is a technique peculiar to television, and prior knowledge of this will enable the designer to think in the correct terms immediately, as well as know how to use the great resources of the television cameras to produce the effects he wishes to convey.

It will be seen that the wider the designer's experience, the more fully can he personally cope with his job. But proficiency in any single one of the foregoing categories of craftsmanship is not sufficient by itself for the multiplicity of demands which are bound to be made upon a designer. Apart from the range of experience discussed above, the designer will need to have at his finger-tips a range of techniques and styles of drawing which are applicable for television, and which he can produce without wasting time in experiment and research. In television work, time is very valuable. The designer cannot afford to wait for inspiration; he must depend upon his background experience to produce graphics at short notice and of high quality.

A further factor is that each programme is governed by a budget, often very limited, therefore the designer cannot afford to be extravagant in his choice of materials; yet he must not allow limited budgets to blinker his design ability. This could make the programme look a compromise between extravagant ideas and a limited budget and destroy the total appearance of the programme. Good design does not cost any more money; many of the best graphics are the result of limited facilities and the designer's ability to channel his experience and ingenuity to create the best possible results. Peripheral factors should never be allowed to impinge upon the designer's good taste and professional standards of design.

CAPTION SIZES AND CUT-OFF

In designing graphics the artist must consider the inherent limitations and problems presented by the medium of television. The most important is that all designs (which are called captions) must be kept to a 4:3 ratio. The standard accepted size is 12 in. × 9 in. and if larger captions are required, 16 in. × 12 in. or 24 in. × 18 in. Given this basic size, however, the problem of cut-off must be faced. It will be observed that the shape of a television screen does not contain any straight lines; they are all curved. This means that it is not possible to transmit the total area of a 12 in. × 9 in. caption, and the areas not included on the screen are said to be 'in cut-off'. There is a set of standards for cut-off, which depend upon the optimum performance of the receiver and the correct line-up by the cameraman in the studio. The average cut-off B is less than the minimum required for complete safety; it is better to allow at least as much cut-off as C, giving a viewing area of 11 in. × 8·25 in. The designer produces captions in order to give visual information to the viewer, and not only is it humiliating to see his work mutilated, with important information hidden by the edge of the screen, but it is also bad design, resulting in severe reduction of the visual value of the caption. Although a slight distortion occurs at the edge of the screen, the designer should always include the cut-off areas in his design, but never place any important information near the edges of the caption.

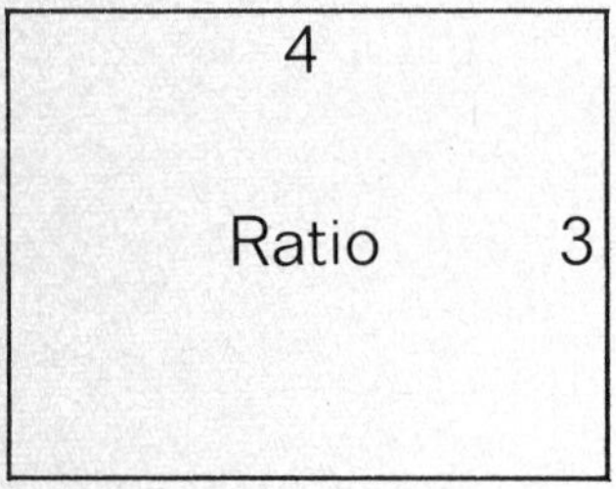

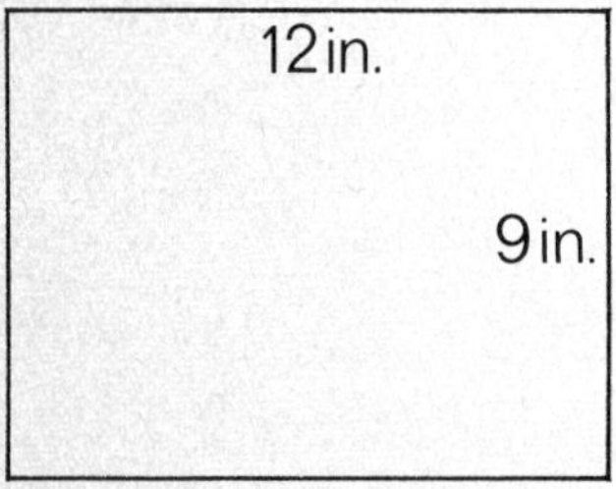

All television images correspond to the ratio of the TV tube, which has a ratio of 4:3. The standard size for TV captions is 12 in. × 9 in., this being a convenient size for producing most types of titling, diagrams, illustrations and photographs. If larger sizes are used, they should always be in the same ratio, otherwise vital information would be lost when transmitted. It is very rare to use captions smaller than 12 in. × 9 in.

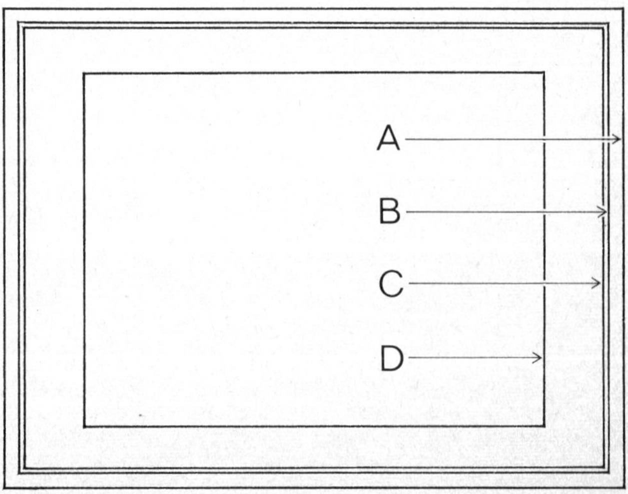

Due to the nature of the electronic image of television, a certain proportion of any caption is lost during the process of recording and transmitting. This lost area is called 'cut-off' and there are recognized standards for this.
A is the 12 in. × 9 in. caption.
B is the minimum cut-off, giving a caption area of 11·4 in. × 8·55 in.
C is cut-off to allow for the over-setting of TV receivers, giving an area of 11 in. × 8·25 in.
D is the safe cut-off area for lettering and titling, which is an area of 8·65 in. × 6·5 in. in the centre of the caption.

For titling it is necessary to allow a larger cut-off area for complete safety, which leaves an area indicated at D of 8·65 in. × 6·5 in. in the centre of the caption. There is nothing more infuriating than having part of a word hidden or cramped, and if the horizontal width control has been overset, this is likely to occur quite frequently. The designer must make full allowance for this and observe the safe titling area very closely. A convenient figure to allow when producing titles is approximately a 1½-in. margin all round a caption. This is a slight over-estimate, but it errs on the correct side. Another method is to cut an area of 9 in. × 6 in. from the centre of a 12 in. × 9 in. caption card. This gives a frame which can be placed on top of a caption card being used for titling, thus eliminating any possibility of overstepping the cut-off margin.

It must be remembered that the cut-off on large captions is in the same proportion as a 12 in. × 9 in. caption, that is, the same proportion of the dimensions of the caption.

SCALE

Because the television picture is made up of series of horizontal lines, there is always a slight loss of definition when the original is compared with the receiver image. Bearing this in mind, the designer must always make his designs bolder, more precise and clearer than normal reproduction standards demand. This will certainly affect the scale of the image in relation to the screen size. The scale must make full use of the 12 in. × 9 in. area, but the design must not be too small or too cramped. This, of course, always includes the observance of the rules of cut-off. The designer must choose the scale of design which will be most appropriate to the subject, simplifying or arranging it on two or three captions if the point to be illustrated is confused and complicated. All drawings or lettering must be of a bold nature; fine spidery lines are not acceptable.

TONE

One of the most important aspects of tone in relation to television is that the designer never produces drawings or diagrams on white paper or card. The accepted colour is a light grey or sometimes a medium grey. A black line drawing on white card has too much contrast; the white areas impinge on the black line and it presents problems for the vision-quality engineers ('racks'), whose job it is to control the electronic picture quality in order to obtain the best possible television image for transmission. The same drawing on a grey caption card would be of the required tone quality.

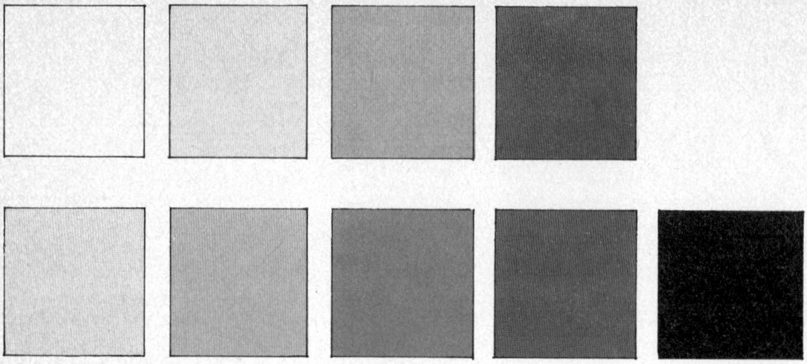

All caption images for television must have a slightly restricted tone range, so that the electronic engineers can obtain the best possible picture quality for transmission. The tones must be well defined and it is very rare to produce any artwork on white card, because this gives too contrasty an image. All captions are produced on grey or black card. If there are largish areas of white in a caption, the darkest tone should be a dark grey, but if there are large areas of black, light grey should be the lightest tone. The exception to the rule is white lettering or drawing on black card.

The tone range for television must be closely observed. The standard is from almost white to almost black. If the range starts at white (only small areas can be tolerated) then the other end is dark grey, with perhaps some black. When black is one end of the scale, the other is light grey. The tone range is always slightly reduced, because the electronic system naturally increases the contrast and range. Also the engineer can change the tonal balance of an image so long as it is restricted to the correct range. This is a general rule, and though it is possible to extend the range at times, some experiment is necessary before transmission. Tones used for television must not be too close, otherwise there is a danger of their merging into one, particularly at the black end of the range, so that the information can no longer be identified. Subtle drawing in shadow areas may not register on a television receiver.

Successive captions must be roughly of the same tone range. A cut or mix from a very light caption to a dark caption can result firstly in a loss of detail in the dark caption, and secondly could result in an undesired adjustment of the tone range by electronic means, while the caption is 'on air'. The converse is also true. The designer must, as far as possible, make allowances for this, so that the racks engineer has sufficient time to adjust the electronic balance of the television image before it is transmitted.

There are several other fields in which tone has to be very carefully considered, such as lettering, hand-drawn captions, photographs and animations; these will be considered in the relevant chapters.

COLOUR

Although colour is being transmitted by the television companies, the majority of viewers do not have colour receivers yet; they are watching a monochrome screen. Therefore, all the colour pictures transmitted must register fully in black and white, otherwise there will be a loss of information. This can present some problems for the designer; a design may be very good when viewed in colour, but it may not translate into black and white because the tones are too close. The tones reproduced by television are not the same as those seen by the eye. Certain colours appear darker or lighter on monochrome television. When producing designs in colour for colour transmission, the designer must use fairly strong colours, as soft, subtle pastel colours do not reproduce too well, and their black-and-white equivalents do not have sufficient contrast.

A general rule for colour to black-and-white conversion is:

COLOUR	CORRESPONDS TO
yellow	light grey
orange	light to mid grey
red	mid grey (sometimes dark)
purple	mid to dark grey
green	mid to dark grey
blue	dark grey
deep blue	almost black

There are some slight variations in the colour/black-and-white reproduction, according to the type and quality of the television equipment in the studios. Some quick comparison tests by the designer are advisable, so that he may be certain of all the monochrome equivalents of the colours he uses in his designs.

2 Lettering

The production of lettering for titles and credits is probably one of the most important and demanding aspects of the designer's work. Almost every transmitted programme requires introductory and closing credits, and some need additional lettered captions within the programme. The demand is so great that the designer must be not only very proficient in producing layout designs quickly, but actually able to letter the captions himself.

With many forms of lettering and production the designer merely designs: the physical make-up is left to other people. This may mean redesigning, because of the impracticabilities of the original design and maybe the designer's own lack of experience and proficiency. In television this cannot be tolerated; the designer must always work within the limitations imposed by the medium, and he must learn them all so that they become an unconscious extension of his normal design-processes. So much time can be wasted producing designs which are not acceptable for transmission, because some letters are in cut-off, the type style chosen does not register clearly on the screen, or it is not appropriate for the subject of the programme. Speed is essential, but it must be subordinate to the standard and quality of design and interpretation.

MECHANICAL LETTERING

Since speed is important for titling, most of the lettering for television is produced by mechanical means, either hot-press or instant rub-down lettering such as Letraset or Norma Letterpress. Both methods are quick and efficient, and give almost identical results when transmitted, but the costs are not comparable. The hot-press equipment is expensive and occupies much space, whereas Letraset is cheap to use and has a much wider range of type faces.

In the hot-press method the printing is from heated type on to card, with a pigmented foil placed between the type and the card. This produces a slightly indented lettering of good quality and strength, except that at times it needs retouching because the edges can be fuzzy or the foil fails to adhere properly. Not only does the press cost over

The lettering on the opposite page shows clearly how the structure of the TV image causes a loss of definition. This is slightly exaggerated here because only a portion of the complete image has been used, but it is obvious that this loss of definition must be considered when choosing type faces for titling. It means that all lettering must be reasonably bold and not too condensed or decorative, otherwise it will be unintelligible when transmitted. The horizontal lines are the means by which a television image is constructed, which is an electronic beam scanning the front surface of the TV tube, 405 or 625 times every $\frac{1}{25}$ second. If you examine a television picture very closely you can see these lines, or raster, very clearly.

One of the ways of producing mechanical lettering is with a hot-press machine. This uses metal type which is set in the normal way, clamped into the machine and then heated. A caption card is placed on the baseboard, and a pigmented foil laid on top. The lever is pulled down, thus imprinting the type on the foil and leaving white or black lettering adhering to the surface of the card. It is a combination of pressure and heat which prints the lettering.

£250, but it is necessary to have a fount of type for each face and in a wide range of point sizes. Normal founders' type can be used, but owing to its comparative softness, particularly when heated, it soon loses its crispness of line. For longer-wearing properties it is necessary to use brass type, which is more expensive. One advantage of the hot-press system is that several copies of a title can be made in considerably less time than it takes to Letraset the same number. Another is the ease of producing roller captions (see p. 21), both horizontal and vertical. The foils normally used are black and white, although if tone lettering is required, blue, yellow and green are suitable.

Letraset and Letterpress are produced in approximately sixty styles, some with variations of weight, bringing the total to about ninety. Altogether there are about six hundred sheets to choose from, ranging in size from 150 mm. (about 6 in.) to 8 point. Some of these styles are not suitable for television, because they do not reproduce well on the screen. Not only is Letraset economical, but it is very easy to use, and with practice is very quick to apply. The letters, which are self-adhesive, are printed on a clear plastic sheet. To apply, place the sheet in the required position, lightly press down the letter with a finger, then rub all over the letter gently with a biro, pencil or end of a brush. The sheet should then be lifted carefully (to avoid breaking the letter), and the applied letter pressed down with a finger. Letraset can be applied to a great many surfaces such as card, board, glossy and matt photographs, cel (triacetate), glass and textured surfaces.

There is a vast range of instant rub-down lettering available, both in different type faces and all the various point sizes, marketed under the brand names Letraset and Normatype. These are extremely convenient for the TV designer, because they can be applied on artwork and captions very quickly, and if mistakes are made, can be removed very easily. This example shows just a small part of the wide range available.

ABCDEFGHIJKLMNOP
abcdefghijklmnopqrstuvwx

ABCDEFGHIJKLMNOPQRSTUVWX
abcdefghijklmnopqrstuvwxyz 123456

ABCDEFGHIJKLMNOPQRSTUVWXYZ&?!;:' abcdefghijklmnopqrs

ABCDEFGHIJKL
NOPQRSTUVWX

LMNOPQRSTUVWXYZ
abcdefghijklmnopqrstuvwxyz,.:;- £$1234567890

ABCDEFGHIJKLMNOPQ
abcdefghijklmnopqrstuvwxyz123

Some styles of lettering are not suitable for television, because there are features of their design which make them unreadable. The faces shown (*top to bottom*), Modern No. 20, Futura Light, Compacta Light, Romantiques No. 5, Palace Script and Old English are typical examples of lettering which should be avoided. There is, in most cases, another type face which has a similar character or style, and which will register clearly on the screen. Legibility is always the first consideration when choosing lettering, particularly for programme titles.

Styles of lettering

Many styles of lettering are suitable for television titles, although some should be viewed before transmission, particularly if they are intended for superimposition on film (telecine) which is rather light in tone. The letter styles not recommended are the exaggerated Modern Roman types, extra light letters, very condensed letters, very decorated styles, copper-plate scripts and some black-letter styles.

Although Modern Romans are variations of the original hand-carved Trajan Roman, which is not only a beautiful letter form but also suitable for television, many cannot be used for television. Television is created by a beam scanning a tube with 405 or 625 lines every twenty-fifth of a second. These lines are called the screen raster. The hairlines and fine serifs of the Modern Romans nearly always coincide with the raster and therefore a vital part of the lettering disappears; only the thick strokes would be visible. A similar effect can be observed with very light letters. In most cases the slimness of the strokes in relation to the point size causes illegibility, and if the point size is increased to gain sufficient strength of line, the letters appear too

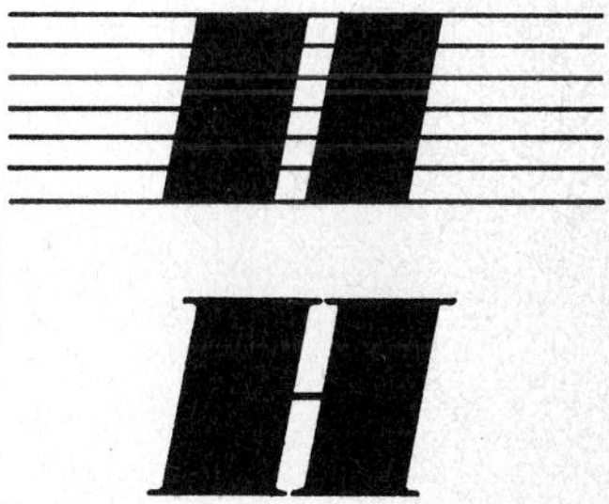

Many of the modern derivatives of the classical Roman lettering are totally unsuitable for television, because they generally have exaggerated variations in weight and line. They look clumsy with their thick, slab-like verticals and extremely fine horizontals when printed on paper, but on the TV screen the thin hairlines and serifs disappear in the screen raster, thus destroying the legibility and character of the letters. Always avoid using styles like this.

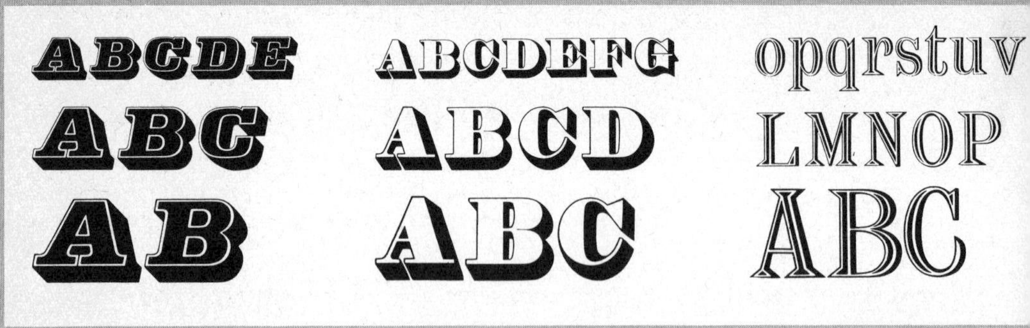

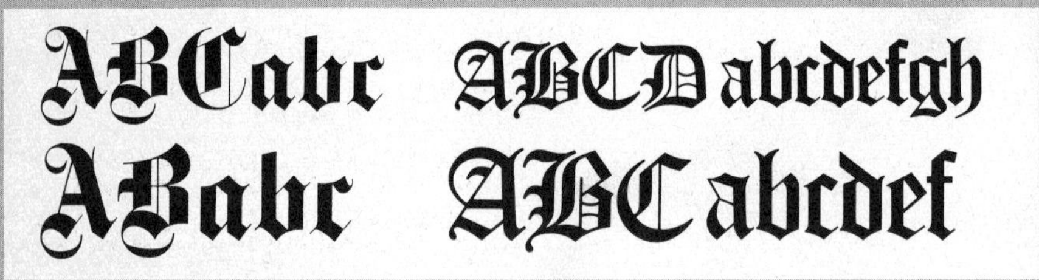

These are further examples of type faces to be avoided if possible. They are (*top*) Slimblack, Compacta Condensed; (*centre*) Profil, Thorne Shaded, Chisel; (*bottom*) Old English and Gothique. All of them have features in their design which tend to disappear, fill in or strobe when on the screen.

large. Only very short words can be produced, unless the screen is to be filled from edge to edge, when the result would look very cramped.

Condensed letters tend to have rather slim strokes, or if their weight is heavier, the inside shapes tend to fill in, particularly in lower-case letters. The normal spacing of these letters is too tight for television, and if the letters are spaced, the words and titles lose their unity. With most decorated letters, the strokes are heavy and include delicate white line decoration. This generally fills in or shimmers, making the letters look ungainly. The proportions of Profil, Thorne Shaded and Fry's Ornamented make them doubtful starters even if they were not decorated. Although Chisel is a slimmer face, the decoration makes it unsuitable. Copper-plate scripts can be used sometimes in their very bold forms, but the normal scripts like Palace Script and Marina are most disturbing on the screen. The extremely fine strokes either do not register or they shimmer. Black-letter styles such as English Script or Saxon Black suffer in the same way as decorated and copper-plate letters, and in addition the inside shapes are inclined to fill in.

SANS SERIF
sans serif

ABCDEFGHIJKLMNOPQ
abcdefghijklmnopqrs

ABCDEFGHIJKLMNO
abcdefghijklmnopqrst

ABCDEFGHIJKLMN
abcdefghijklmnopqrs

ABCDEFGHIJKLMNOP
abcdefghijklmnopqrstuv

ABCDEFGHIJKMNO
abcdefghijklmnopqrs

ABCDEFGHIJKLMNOPQ
abcdefghijklmnopqrstuv

ABCDEFGHIJKLMNOPQRSTUVWXYZ&?!;:-
abcdefghijklmnopqrstuvwxyz12345678

ABCDEFGHIJKLMNOPQ
abcdefghijklmnopqrstu

ABCDEFGHIJKLMNOP
abcdefghijklmnopqrstuv

ABCDEFGHIJKLMN
OPQRSTUVWXYZ

ABCDEFGHIJKLM
OPQRSTUVWXY

ABCDEFGHIJKLMNOPQRTSUV
abcdefghijklmnopqrstuvwx

Other letter styles are not recommended, because of their inferior design. These are often modern type faces, but their design is inconsistent; some letters are ugly or details, such as serifs and weight of stroke, are unsuitable for good television design. Any exaggerated type style should be used only with discrimination, and if possible tested before transmission.

Although some styles of lettering and type are not suitable for television, there are many styles left from which to make a choice. The choice is dictated by the style and content of the programme, as is the layout.

One of the most suitable and commonly used styles are the sans serif group of letters. Their strength of line, character and shape reproduce very well on television, and the variations available make them very appropriate for many categories of programmes. They have sufficient boldness for short titles, as well as finesse when used in mass. The smaller point sizes and lower-case letters are very clear, and with discretion they can be blended with other styles. Faces such as Grotesque 9 or 216, Folio, Univers, Compacta, Din, Futura, Helvetica, Microgramma and Futura Display are particularly suitable.

The sans serif letters are the most commonly used and apposite form for television. They have strength of line, good weight and clarity, as well as variation of character, and they are appropriate for a wide range of programmes. Styles such as Grotesque 9, 215 or 216, Folio, Univers, Compacta, Futura, Helvetica, Microgramma and Futura Display are very suitable.

(*Above left*) When lettering is to be superimposed on another picture source it is always produced as white lettering on a black card. This method is also used for word captions and it gives good clarity.

(*Above right*) When white letters on black are superimposed on another picture source, the black is adjusted so that it disappears, and only the white letters are visible on the other image. Care must be taken to ensure that the white letters are superimposed on a medium to dark tone area of the picture source.

(*Right*) In some cases there is not a dark area in a picture which will allow superimposition of white letters. This is overcome by using a normal white on black caption; the letters are phase-reversed so that they become black, and then inlayed on to the picture source.

Superimposition

Much of the titling used for television is superimposed on to another picture source, such as telecine, studio shots or another graphic, and appear as white letters on the screen. A superimposed title must be produced by using white lettering on a black caption card. The black is electronically adjusted ('crushed') so that when it is electronically superimposed on another picture source, the black disappears, leaving only the white letters and the other visual. This method can also be used for diagrams, symbols and line drawings. Even when superimposing on colour television pictures a white on black caption is used, and if necessary colour can be injected into the white letters with an electronic synthesizer.

Sometimes the style of the lettering is determined by the quality of the picture source on which it is being superimposed. If this is rather light in tone, then it is necessary to use a fairly bold type. At other times the titles will have to be accurately positioned within a specific area of the screen, perhaps to avoid a light area or so as not to obscure a particular object or action. With 'supers' (see p. 29) the background shows through the lettering very lightly, and to

When lettering is not superimposed it is possible to use black letters on light-grey card, white and black letters on medium-grey card, or black and white letters on light-grey and black card (split-screen arrangement). This can give both variety and emphasis to word captions.

improve the strength and quality of the titles it is possible to use a facility called inlay. A normal white on black caption is used, and by electronic means the titles are 'punched into' the background, so that they are opaque white. If the telecine picture is very light, it is possible to produce a reversed image with inlay so that the lettering becomes black. This particular facility dictates that the style of lettering must not be too light in weight and each letter must be well spaced.

If superimposed titles are not required, then black letters on light- or medium-grey caption card can be used, or perhaps white and black letters on medium-grey card. A combination of both black and grey card is acceptable for some titles, and also for charts used in news bulletins and educational programmes.

Roller captions

When the designer receives his brief to produce titles for a programme, he must determine with the producer or director whether they have to be superimposed or not, and whether they are 12 in. × 9 in. captions or a roller caption.

Roller captions are titles and credits produced on a length of black paper, which is fitted into a machine that moves the lettering on the screen, either vertically or horizontally. The machine has rollers at the top and bottom and a viewing aperture in between. The rollers are coupled to a variable-speed electric motor, and the paper is transferred from one

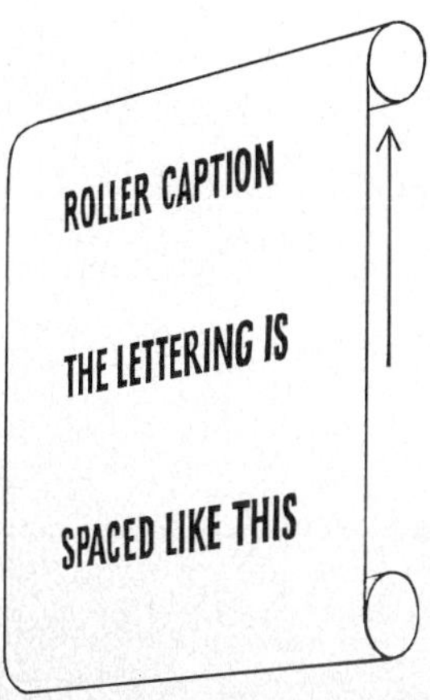

(*Below left*) Instead of having static word captions, words or lines of lettering can move up or across the screen. This is achieved with a roller caption, a length of black paper, 12 in. wide, which is fitted on rollers. The lines of lettering must be spaced well apart, to aid reading and so that there are not too many lines on the screen at once.

(*Left*) The roller is fitted into a machine, which has a variable-speed electric motor, and the paper is transferred from the bottom to the top roller, while the camera shoots the area of the aperture. The danger with roller captions is that often they are run too fast, and it is very difficult to read the words.

roller to the other while the camera shoots the area of the aperture. When transmitted, the lettering can move upwards so that there are as many as five lines of lettering on the screen at once, or just one line at a time. The roller machine can also be swivelled so that the lettering moves across the screen from right to left, the normal direction of reading.

Spacing

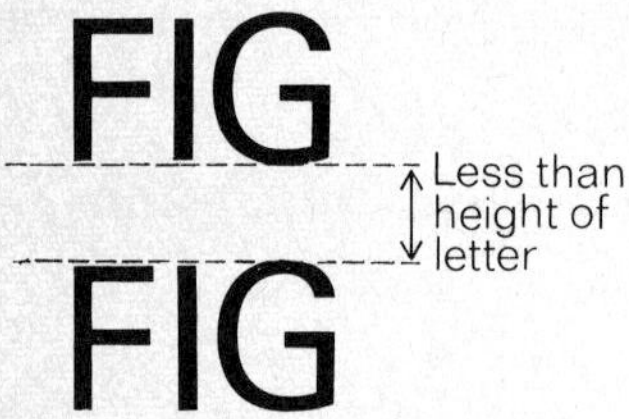

For word captions and titling the linear spacing between lines should be less than the vertical height of the type used, but not smaller than a quarter of the height of the letters.

The first consideration in spacing is the observance of the safe cut-off areas for titling (see p. 10). With the normal 12 in. × 9 in. caption, only the centre area of $8\frac{3}{4}$ in. × $6\frac{1}{2}$ in. must be used. Any letters outside this area are liable to be masked by the edge of the screen. The observance of cut-off is as important as the need for correct margins in book design. To help remember the correct cut-off, as we have seen, it is best to cut a section from the centre of a 12 in. × 9 in. caption card, which corresponds to the safe cut-off area, and place this on the caption being lettered, thus preventing accidental overstepping of the margin.

The linear spacing of lettering is very important. The distance between lines should not be more than the height of the letters being used, i.e. if $\frac{1}{2}$-in. letters are being used, then the distance between should not be more than $\frac{1}{2}$ in., preferably just a fraction less. Also the distance between the lines should not be less than a quarter the height of the letters. Too little linear space makes the lettering look cramped. There are always exceptions to these rules, but this is only for very special title designs.

It is essential that all lines of lettering be parallel to each other, as well as parallel to the bottom of the caption. A useful aid to ensure parallel lettering is to make a holder as follows: Sellotape a 12 in. × 9 in. grey caption card (A) to another grey card 14 in. × 14 in. (B). The top edge of card A can be marked to show the cut-off limits and the centre (6 in.). The caption to be lettered should be marked at the edges to show the position of the line of lettering; then slide it between A and B, so that the marks line up with the top edge of card A. The line of lettering should now be straight so long as the

An aid to ensure correct spacing of lines of lettering, and to make sure they are parallel is this holder. A 12 in. × 9 in. card (A) is Sellotaped to a 14 in. × 14 in. card (B) and marked as shown. The caption is then marked at the sides, slipped between A and B, and the marks lined up with the top edge of A.

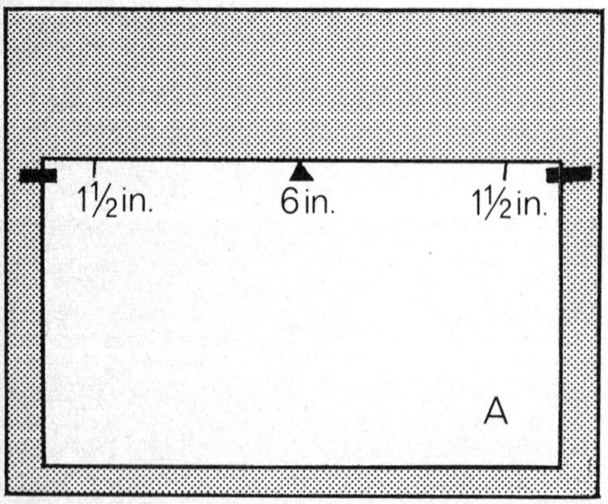

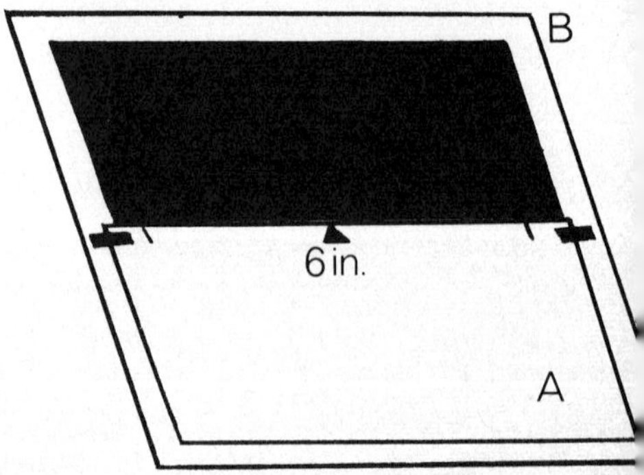

measurements at the side are correct. When more than one line of lettering is required, a simple method of calculating the distance between each line, and how far from the bottom of the caption it needs to be, is to make a table for each size of lettering.

For letters 3/4 in. high (60 point) the distance between the bottom of one line of lettering and the bottom of the next line should be approximately 1·2 in. According to the number of lines of lettering, the bottom margin will vary. This table gives the distance between the bottom edge of the caption and the bottom of the lowest line of lettering for varying numbers of lines.

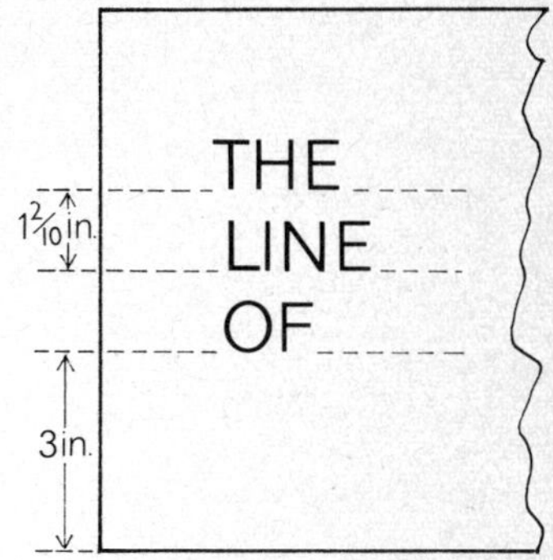

When 3/4 in. letters are used, the distance between the bottom of one line and the next should be approximately 1·2 in. According to the number of lines needed, the table indicates the size of the bottom margin.

3/4-IN. LETTERS 60 POINT

number of lines	margin at bottom
1	4·1 in.
2	3·7 in.
3	3 in.
4	2·3 in.
5	1·75 in.

Measure and mark the bottom margin, then mark every 1·2 in. upwards, according to the number of lines of lettering. Six lines of 60-point lettering are not recommended, because there is insufficient space on one caption and some of the lettering will be in cut-off. For 1/2-in. letters (48–36 point) the distance between the bottom of each line of lettering is 1 in. and the bottom margins are:

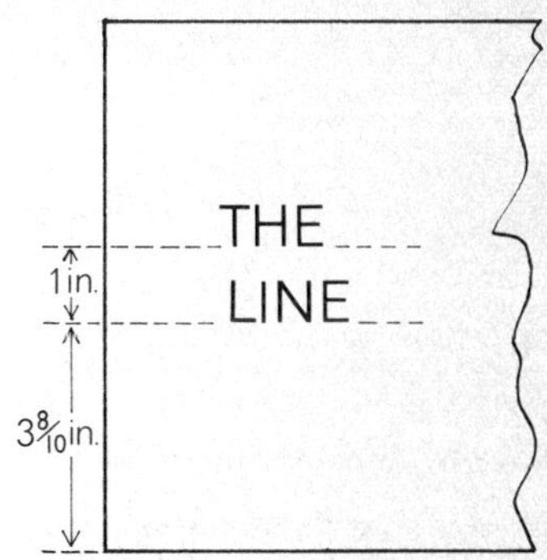

For 1/2-in. letters the linear spacing is approximately 1 in. and the table shows the bottom margin.

1/2-IN. LETTERS 48–36 POINT

number of lines	margin at bottom
1	4·3 in.
2	3·8 in.
3	3·3 in.
4	2·8 in.
5	2·3 in.
6	1·8 in.

When there is a mixture of type sizes, the linear spacing should be done visually, then measured up in order to obtain parallel lines. When using Letraset, it is convenient to line up the thin rule printed on the sheet (just below the letters) with the top edge of card A, thus ensuring accurate placement of the letters.

The dotted lines printed on Letraset and Normatype sheets can be lined up with the top edge of card A, when different-sized lettering is used on one caption. It ensures correct alignment of the letters.

Linear spacing for roller captions must be more variable. When lines of lettering are static, up to six lines are acceptable on one caption, but on vertical rollers, the lines need to be farther apart with not more than four lines on the screen at once. If more lines are presented, the viewer encounters difficulty in reading unless the roller is run at a very slow speed.

When one line at a time is required on the screen, then there must be at least $8\frac{1}{2}$ in. of space between each line, so that the top line is just disappearing when the next line appears. When the roller is horizontal the lettering, being spaced sideways, can be in closer groups, preferably in single lines. Two lines can be tolerated, but more is very difficult to read.

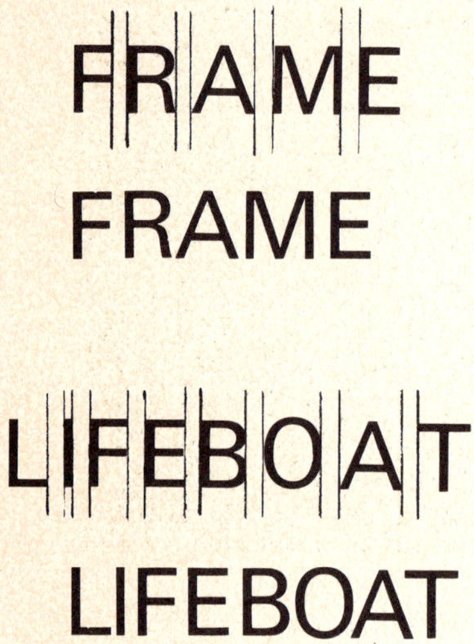

These two examples show that the spacing of letters cannot be done mechanically. Because the letters vary in width, all spacing must be done visually, otherwise the words would appear disjointed. Full letters must be placed close together, whereas slim vertical letters need more space around them.

Letter spacing

The spacing of letters and of words is two totally different problems. There are no mechanical aids for spacing letters; it must be done visually. The most important point to remember is the amount of 'air' – the space in and around the letter – which governs the letter spacing. Some letters, by their nature, demand more space than others and have to be balanced with slimmer letters. This varies from one type face to another.

When letters of natural spaciousness such as A, C, L, M, O, T, V and W are combined with slimmer letters, more spacing is allowed for the latter than the former. In the word 'FRAME' there is exactly the same amount of space between each letter, but the F and R and the M and E are too closely spaced in comparison with the space between the R and A and the A and M. This space cannot be reduced, so more spacing is required between FR and ME, so that it is now visually spaced with regard to the 'air' in and around the letters, and the word is uniformly spaced. Two other examples are shown, the first set badly spaced and the second set correctly spaced. Letters of natural fullness can be spaced almost touching each other, whereas thin letters require wider spacing to avoid looking cramped.

When several lines of lettering are arranged to line up on one side, it is necessary to make allowances for spacious letters, such as A. These letters must overstep the margin a little, otherwise the line appears to curve inwards, owing to the additional air space of the A.

With lower-case letters the difference in the width of slim and spacious letters is not so pronounced, but many of the points made above are still applicable. Lower-case letters are generally used when phrases, sentences and state-

When lines of lettering are aligned one side, allowances must be made for letters of a full or open nature. The example on the left has been lined up vertically, but the second line appears to be indented, owing to air space round the A. The right-hand example appears to have the correct vertical alignment, because the A oversteps the margin fractionally.

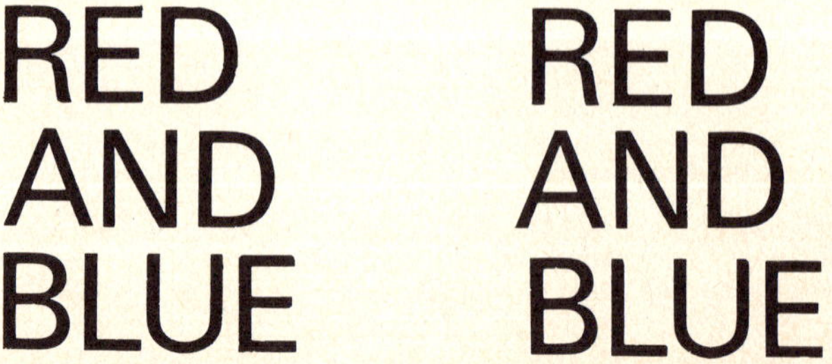

Even in the mid-twentieth century the city of Fez still preserves craft techniques and intellectual traditions dating back to before 900a.d. Here,until a few years ago,Islamic alchemists were still practising their traditional arts. In this film we show this aspect of mediaeval science in its proper setting.

We gratefully acknowledge the co-operation of our colleagues in Morocco,especially those at the ancient Qarawiyyin University

ments are required on one caption. Obviously because they are smaller than capital letters, many more words can be accommodated on one caption, although more than five lines should be avoided. When more than five lines are used, the point size of the lettering needs to be smaller than 48 point or 36 point (the ideal sizes) so that the texture of a mass of words becomes a little confusing to read. Lower-case letters should not be given too much space, for then the words tend to become disjointed. They are normally used in conjunction with their appropriate capitals, although titles are sometimes produced with only lower-case letters. This should be used with discrimination; it is only appropriate for a limited number of programmes and subjects and it can look so ungainly in comparison with upper-case titles, mainly because large point sizes must be used to give sufficient scale and weight. The real merit of lower-case

When a lot of information is needed on a caption, there is a temptation to reduce the size of the lettering to fit it all in. This is a mistake: the example has too many lines, and when transmitted the words would be extremely difficult to read. This number of words should be split into two captions.

Upper &
Lower Case

When there are a lot of words on a caption, it is always advisable to use upper- and lower-case letters. They are easier to read, we see them daily in newspapers, books, etc. and they are better for introducing new words.

letters is familiarity – viewers see them in this form in newspapers, books and magazines and read them naturally. Also for educational programmes, when new words are shown on the screen, pupils absorb the lower-case form more readily.

Centring

Even though the centred layout is so widely used, it presents more problems than nearly all other forms of layout, except with the hot-press method of lettering. With hot-press letters, the type is set up and can be physically moved to any position before printing. With Letraset and hand-drawn letters it is not possible to print a whole word or line at once, only single letters. Therefore the first letter applied must be exactly in the correct position, otherwise the word will not be centred on the caption.

The normal method of centring is to count the number of letters in the word or line to determine the centre letter. Each space between words is counted as one letter. If the number is even, say four, then the centre is between the second and third letter; if it is five then the centre is the third letter. However, remembering the variation in fullness of the letters, the numerical centre is not necessarily the true centre of the word. The letter I, being less than half the width of most other letters, calls for most adjustment in centring. In the word 'DESIGNERS' the numerical centre is G, but owing to the I the centre of the word is between G and N. When I is contained in both halves of a word, they balance each other and no adjustment is necessary. Other letters, such as M, W and O, are slightly fuller than average, and some allowance must be made for this.

the
space
age

Lower-case letters can be used by themselves for titling in some special cases. The problem is that large sizes must be used to give sufficient weight, and then they tend to look a bit ungainly.

ROAD PAPER

DESIGNERS

DESIGNERS

DIFFERING

The centring of words on a caption can be done visually or by guess-work; however, it is recommended that it should be done by finding the numerical centre of a word by counting the letters. If there is more than one word on a line, the space between words is counted as one letter. The first examples show simple words. In many cases, however, the numerical centre is not the centre of the word because certain letters are very slim and others more spacious. The letter I causes the most problems and the next two examples show firstly the numerical centre and secondly the true centre, which makes allowances for the I. The final word has an I in both halves and therefore the numerical centre is correct in this case.

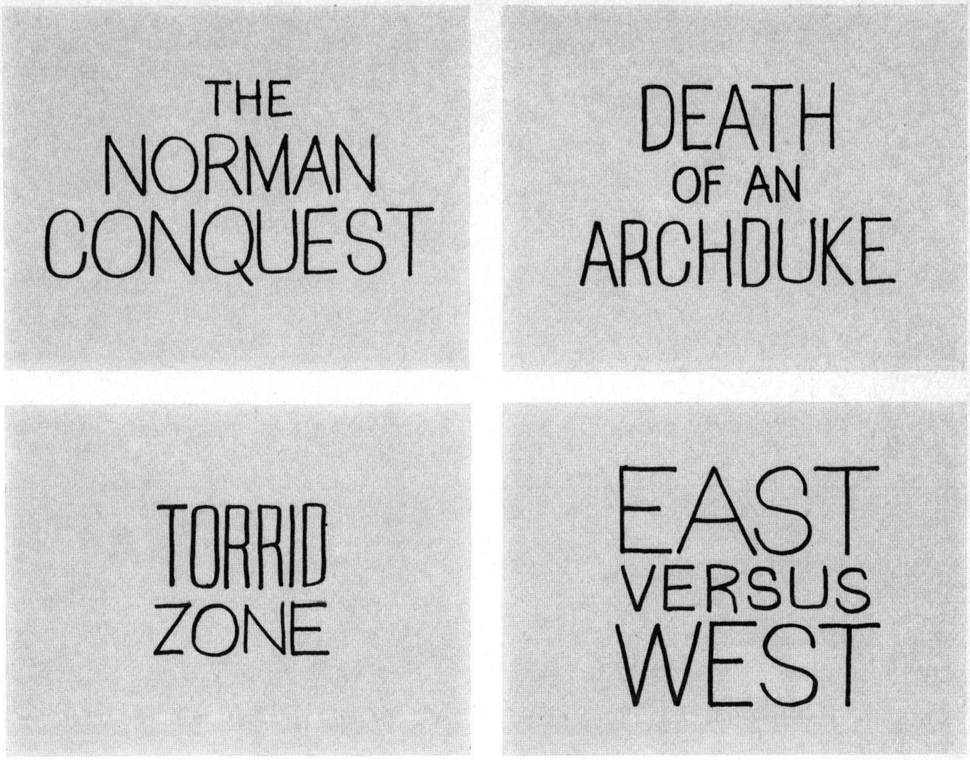

The four layout roughs (*above*) all use a centred arrangement, which in two cases gives a very compact unit of lettering. They also display a good use of different type sizes and styles, which gives a certain emphasis to particular words. The last layout also uses the lettering in an expressive way, where the word 'VERSUS' seems to act as a barrier between the other two words. All the layouts have a clear and compact quality and they are related to the 4:3 ratio of the screen.

When lettering, the first letter to be applied should be the centre letter and then successive letters on each side until the word is completed. To make the problem slightly easier, the holder described earlier can be used, because the centre is marked on card A and the lettering can be applied parallel to the caption.

Layout

Although cut-off imposes certain limitations, the layout of titling can be varied considerably, so long as it is in keeping with the style of the programme. The form most often used is the 'tombstone' arrangement, when the centre of each line of lettering coincides with the centre of the caption (6 in.), giving a symmetrical balance. By using different point sizes and weights of lettering, the layout can be made livelier and the message is conveyed much more quickly. A variation of this style is to make a compact unit of words, by blending two or sometimes three different type faces, and also reducing the linear spacing, so that the letters almost touch. This should be used carefully because of the danger that the words may become unintelligible. Arrangements such as these need careful working out if they are not to look untidy and destroy the sense of the title.

The example 'EAST VERSUS WEST' is more than just a compact arrangement, because it also evokes a feeling of two large masses held apart by a slim barrier. In all these examples, not only are the individual letters related to a unit, but the unit is related to the 4:3 frame. Many arrangements can be conceived, but they would look untidy and uncomfortable if they were not related to the ratio of the screen.

Side alignment of lines of lettering can be very attractive, although, again, it must be used carefully and related to the screen ratio. There is nothing worse than an unbalanced title, caused by insufficient thought on the part of the designer.

The same applies to asymmetrical layouts. The lettering must be conceived as a unit, not as several words divorced from each other. When lettering is integrated with designs or superimposed on film, asymmetrical layouts are often the best solution, and they can be very attractive.

Corrections

The use of white lettering on black card makes corrections of designs a simple matter. If there is a miscalculation when centring lines of lettering, the incorrect line can be repositioned without fear of the alteration showing on transmission. This is due to the electronic 'black crush' facility of television, which eliminates any joins and overlaps on black

The top two layout roughs (*below*) use an asymmetrical arrangement and the words remain a compact unit.

The two lower arrangements have the words lined up on one side; the first is a reasonable solution, but the second is unbalanced and rather messy. All the layouts on this page give some idea of the variety of approaches required for lettering.

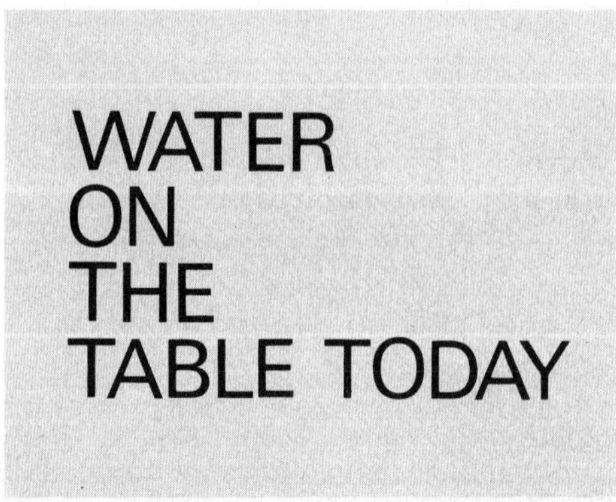

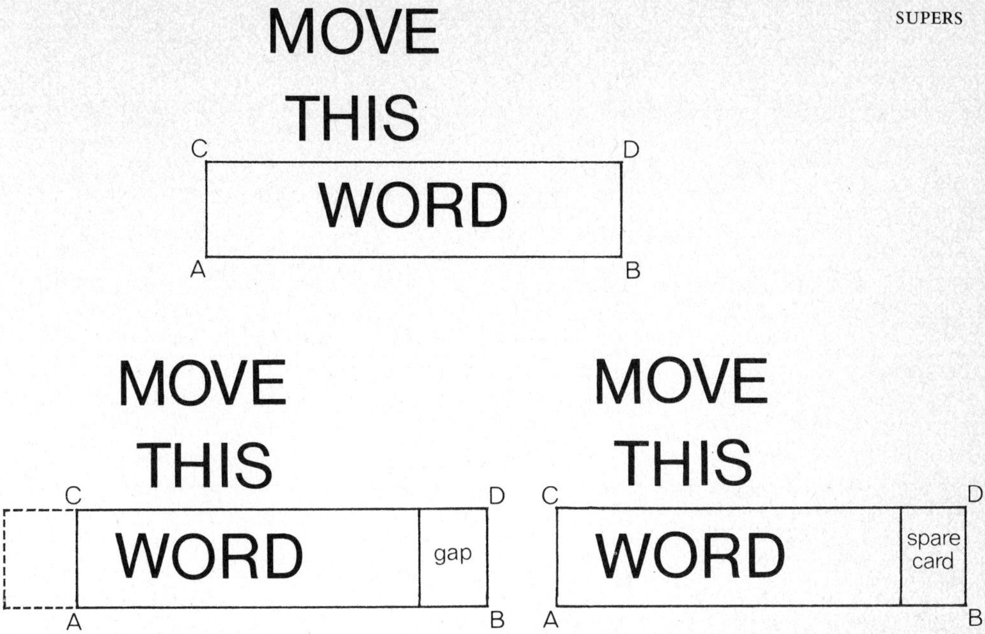

Simple method of re-centring a wrongly placed word or line.

surfaces. These would almost certainly show if grey or toned papers were used.

The method is to make a knife cut just below the incorrect line or word (AB), making sure that it is parallel to the bottom of the caption, and longer than the word itself. Then a second cut is made above the word (CD) parallel to the first. Now by cutting on both sides (AC and BD) the word is movable. Place it in the correct position by moving it to the left so that the spare card is underneath the caption, cut off this card at AC and place it in the gap on the other side. It should fit exactly, and a few pieces of black adhesive tape at the back will secure the word in its new position.

Another method, used when several lines need repositioning, is to cut out each line individually and then stick them on to a new caption card with adhesive or double-sided Sellotape.

Supers

This is the term given to words or names which are superimposed in the lower third of the screen, in order to introduce new words or to name performers and places. These can be superimposed on film, TV camera sources or videotape recordings. Generally the word or line of lettering is placed $1\frac{1}{2}$ in. from the bottom of the caption and centred. When two lines are required, the top line must be $2\frac{3}{8}$ in. from the bottom of the caption, the lower line still at $1\frac{1}{2}$ in. Sometimes it is necessary to position the lettering to the left or right of centre, in order to avoid an important part of the image on the screen.

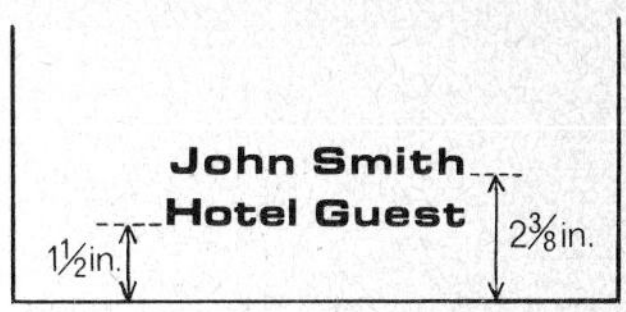

(*Above*) Supers are words or names which are superimposed in the lower third of the screen to identify the picture. They are always white letters on black card. If there is only one line this is positioned 1·5 in. from the bottom of the caption, and if two lines are needed, the top line is approx. 2·2 in. from the bottom.

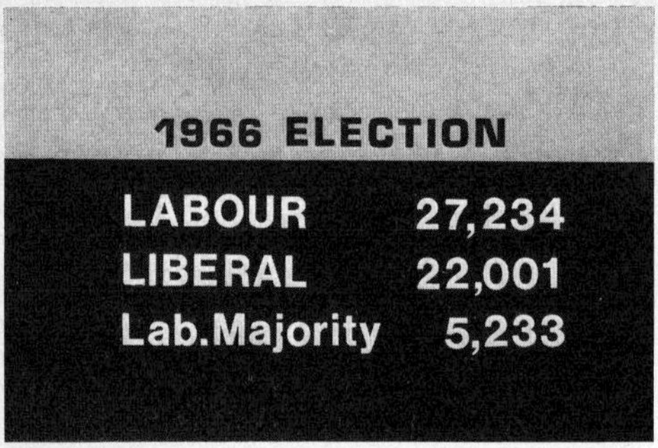

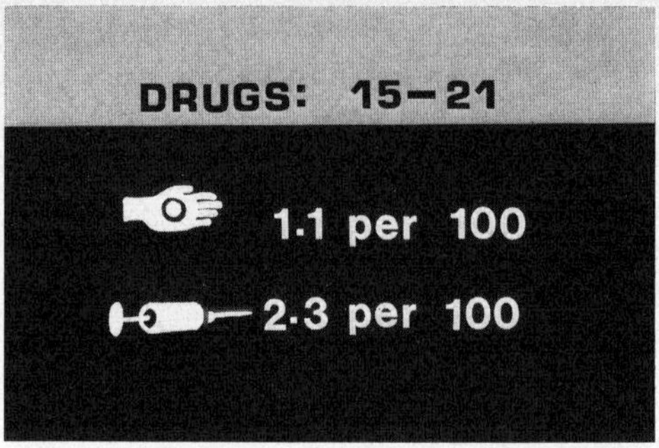

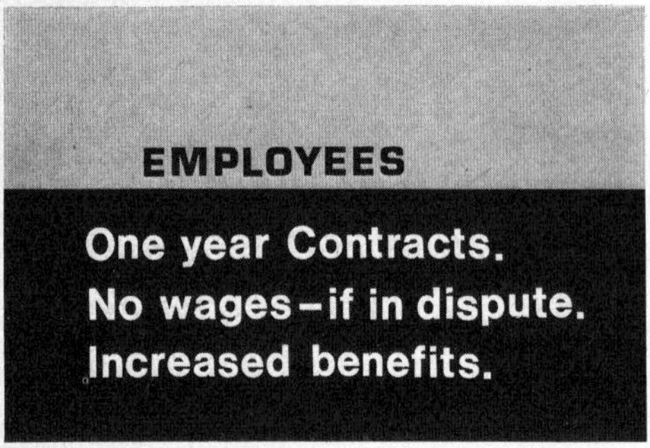

These four captions are typical requirements of news and current affairs programmes. They use a split-screen arrangement which gives emphasis to some of the information and makes the captions more attractive. This arrangement also allows the combination of varied information without causing confusion. Sometimes a simple symbol can be integrated with the lettering and provide additional clarity. It will be noticed that the information is clear and well arranged for maximum legibility.

Sans serif letters in upper and lower case give the clearest and best results, using 48-point letters. The quickest results are achieved by using the holder described on page 22.

Split screen

To make a lettered information caption more varied and interesting, as well as giving some emphasis to specific words, a combined grey and black caption can be used. This gives a split-screen effect, the split being either horizontal or vertical according to the subject and information. A very straightforward presentation is used in the first example, and by using black and white letters the topic title and the information are sufficiently separated. A similar treatment has been used in the next example, but because there were rather a lot of words, capitals and lower-case letters were necessary, to improve the legibility.

In some cases simple symbols are used instead of words, in order to reduce the number of words on the caption, and to make the information more attractive.

The vertical split screen is very useful when comparisons between two sets of information are required, or when two unconnected items are needed on one caption. With a vertical split screen the frame can be divided in half, or one-third and two-thirds, but a horizontal split should never be more than one-third of the height of the screen, otherwise it tends to become very unbalanced.

Presentations like these are valuable for news and current affairs programmes and for educational television, because they give increased clarity to the information.

HAND LETTERING

Hand lettering can be a very useful supplement to the mechanical forms of lettering in providing a variety of styles which are more appropriate to the subject of the programme. It must be stressed at this point that, in general, hand lettering requires considerably more time to execute than mechanical forms, and considerable skill and application are needed to produce hand lettering of high quality. However, the more hand lettering you do the more proficient you become, so that it develops into an extension of normal handwriting.

The basis of many hand-drawn letters is the Trajan chisel letters, a development of handwriting or a stylization of founder's type. When hand lettering is used it must not be a substitute for mechanical lettering; it must have a quality of its own, otherwise there is no point in spending a lot of valuable time on it. Hand lettering has to keep to the same limitations as mechanical lettering, where scale, weight of letter, clarity and weight of line are concerned. Hand lettering tends to have more character if it is not superimposed on another picture source, although it can be very effective when integrated with a design.

The easiest forms of hand lettering are those created by merely writing with a brush in an exaggerated form of handwriting, with the brush held either upright or sideways. Ideally, this lettering should be painted directly without any outline, but this takes a lot of practice. The only aid that is necessary, is to rule the base line in pencil, to make sure that the lettering is parallel to the bottom of the caption.

This form of lettering is extensively used in sports results sequences and in some news presentations; it is not only very fast to produce (particularly if the caption is

Hand lettering is a useful adjunct to mechanical lettering, and can be a quick and evocative form of lettering. At its simplest, hand lettering with a brush is an exaggerated form of handwriting and can be produced almost as quickly. Good hand lettering should not try to imitate mechanical letter forms.

(*Above*) Hand lettering is very appropriate for drama, historical, arts and light entertainment programmes, but the style of letter must be carefully related to the subject. Also the designer must avoid all the factors which make some mechanical letters unsuitable for TV. Styles like these are generally adaptations of type faces.

prepared with some information, and the final details are hand-lettered within seconds of the news breaking) but it also creates an impression of immediacy.

Hand lettering can be very suitable for programme titles, particularly for drama, historical documentaries, arts programmes and even some light entertainment. In all the examples the style has been carefully related to the subject of the programme, and the clarity of the letters has had much consideration.

A useful hint when doing brush lettering is to cut the tip off the brush, because this carries very little paint, has no firmness and can cause unwanted flicks which will require retouching. Also, it is preferable to have the paint fairly fluid so that it flows smoothly off the brush. Paint which is too thick tends to cause ragged edges, thus destroying the fluid character of the letters. This ragged effect can be used to great advantage sometimes, but it must be used discriminately, otherwise it can look like bad lettering.

The special quality of hand lettering is that it has a lot of character and adaptability; the designer must be careful not to destroy this quality and make it too mechanical.

(*Above*) Black Letraset on cel, contact-printed on photo paper, is the simplest form of photographic lettering. It produces white letters on a black ground. The Letaphot machine (*below*) prints letters on photo-sensitive paper, the desired letter being selected by the dial.

PHOTOGRAPHIC LETTERING

There are several methods of producing lettering photographically. The simplest method is to Letraset the required words in black on cel (triacetate) and then contact print it on to matt photographic paper. The lettered cel acts as a negative and produces white letters to the same size on a black ground. These letters and words can be cut out and pasted up on designs and captions. It is advisable to place a sheet of glass on top of the cel, to ensure perfect contact between the cel and the photographic paper. This method requires a darkened room, but it is not necessary to use an enlarger. The paper can be exposed to an ordinary light bulb source for two or three seconds to provide a full print.

If an enlarger is used, make sure the condenser and negative are dust-free, and then expose the paper for approximately ten seconds with the lens set at *f*8. The enlarger method is very useful when several prints of one lettering are required. If a masking frame is used, the lettering can be reproduced in exactly the same position on each print. The prints have perfect white letters and a very rich black background.

Another method is to use a Letaphot or a Varitype machine, with which the operator dials the required letter and it is printed on to photographic paper. Some of these machines print the letters in only one size, while with others the size can be altered. Most of them will print only black letters on a white ground, which can be a severe disadvantage unless it is possible to reverse-phase the caption camera, thus making the lettering white.

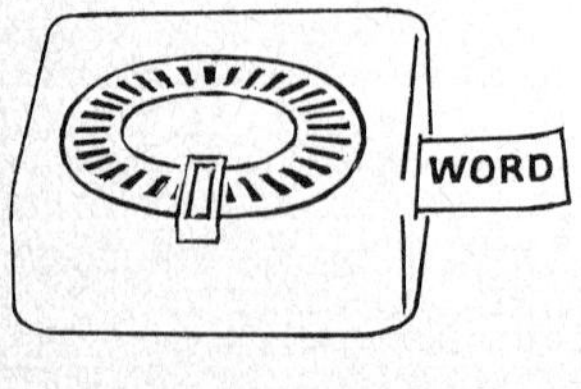

A third method, often used when a particular type style is required which is not available in instant lettering form, is to photograph type specimen sheets on sheet line negative film (5 in. × 4 in.) with a plate camera. After processing, the negative is contact printed on to another sheet of line film, so that when it is printed the letters will be white. Several prints are made from the second line shot in order to obtain sufficient letters. After developing and drying, the letters are cut out, arranged and pasted-up to make the required words. They are normally pasted on to a black caption card or on to designs, and these can be transmitted immediately. If a more permanent caption is required, the paste-up can be rephotographed on line or tone film and a new print made.

The advantage of this method, even though it involves more work, is that unusual type styles can be used, and enlarged or reduced to any size.

Another method of photo-lettering is to photograph type specimen sheets, make several prints, then cut out the required letters and paste them on the design or caption card. If it is copied on line film all the joins and edges will disappear. This method can be very useful when unusual styles or sizes are required.

ELECTRONIC LETTERING

A very recent development in the production of superimposed lettering is the use of a special typewriter, which can print lettering on the screen at high speed.

All the words and information can be stored in advance in a small computer memory bank, and then selected and superimposed instantly, using a simple code. This is used for very high-speed productions which are live transmissions, such as the moon landings.

At present this is not a replacement for programme titling or most of the lettered output of the designer, although it is ideal for news presentation or anything which requires high-speed information. It is a purely mechanical method of lettering, generally typed out by a specially trained typist; no design is involved.

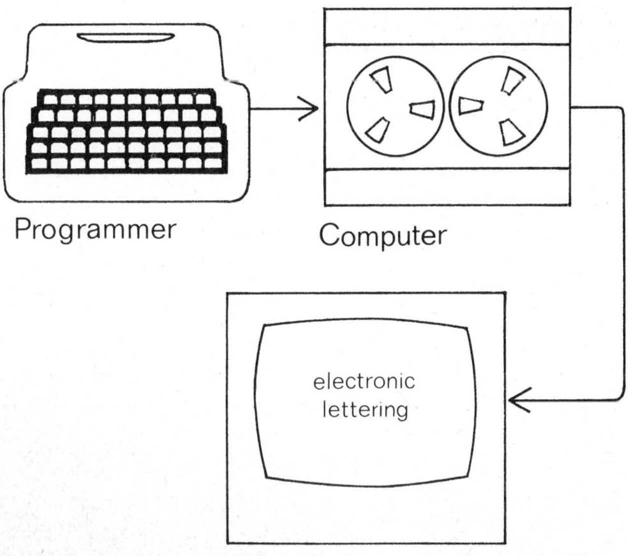

One of the latest developments in instant lettering information is a special typewriter that can print words on the screen in a fraction of a second. These words can also be stored in a memory bank and then selected and superimposed on the screen when required, by using a simple code.

3 Hand-drawn captions

Although much of the design work demanded for television involves lettering for titles and credits, a very important part of the designer's function is the production of hand-drawn captions. These can range from simple line drawings and maps to detailed drawings, illustrations and diagrams. Not only is there a wide range of types of captions, but there is also a demand for a variety of styles and materials, all of which must be suitable for reproduction by television and must not take too long to produce.

The designer must have a wide range of skills and styles at his finger-tips to cope with the many problems. One of the designer's skills is his ability to select and manipulate images which can be very explicit or evocative, and to give the viewer exact details and synthesize complex ideas and moods. In this way the designer can reduce complicated images to basic forms, expressed in terms of a drawing or diagram which will provide clear information for the viewer. In his powers of selection the graphics designer cannot be rivalled even by photography at its most selective. With the growth of educational television at national and local levels, the role of the designer becomes vital in the production of hand-drawn captions to express and emphasize teaching points.

As with all aspects of designing for television, the designer must consider the inherent limitations of the medium. These are the 4:3 ratio, cut-off, the tone range, and the scale of the image. Within the framework of these limitations the designer can be very creative and can use a wide range of techniques and materials. The only proviso is that the subject must remain more important than the method used to illustrate it. Once the designer has received from the producer the information he needs to produce the caption, he must decide upon the style of drawing which will give the best possible delineation, enhance the information, and be attractive and well designed.

MATERIALS

The range of materials at the designer's disposal is very large, though some of course are used only occasionally for special jobs. The stock-in-trade materials are light and

(*Opposite*) This caption is a combination of many of the techniques mentioned in this chapter. It uses elements of hand-drawn imagery, collage, hand tinting, photography and old prints.

medium grey card, black card and 12-sheet board, black fibre-tipped pens, designer's gouache greys Nos. 1 to 5, and a range of colours, black and white paint, double-sided Sellotape, Copydex, oil crayons and mechanical rub-down tints and textures. Additions to these are coloured and textured papers, newspaper, cel, plastic emulsion paint, coloured inks and fixatives; steel rulers, set-squares, ruling compass, french curves, surgical knives, sponges and a range of sable brushes may be needed with these.

Pen, coloured papers, paints and inks are used for both colour and monochrome transmission; it is advisable for the designer to make samples and test them with a television camera, and observe the monochrome equivalents on a TV monitor. These samples can be marked light, mid, dark or very dark grey for future reference. Some colours alter their tonal values when viewed on television; this is due to the type of television tube in the camera. Tests will reveal these differences, which should be noted. Test samples can also be made for textured papers, to make sure they are not too 'noisy' or 'strobe'. Strobing is the technical term for the shimmering effects seen on a monitor and is caused by horizontal lines interfering with the horizontal lines of the raster of the television image.

STYLES

The style of drawing used is always determined by the content of the programme, but also to a certain extent by the choice of materials. A pen cannot reproduce the flowing line of a brush or have the same effect on a textured paper. A map with lettering could not be made on a strongly textured paper, because the lettering would be difficult to read and the caption would be confusing. Legibility and information are more important than personal taste.

Two-dimensional maps

The most straightforward style of drawing is the two-dimensional diagrammatic, which is used in a wide variety

The map (*below left*) of part of the Norwegian coast is very accurate, showing all the indentations of the coast, but it would take a long time to produce. The map on the right has a simplified outline of the coast, which is fairly accurate, and can be produced quickly.

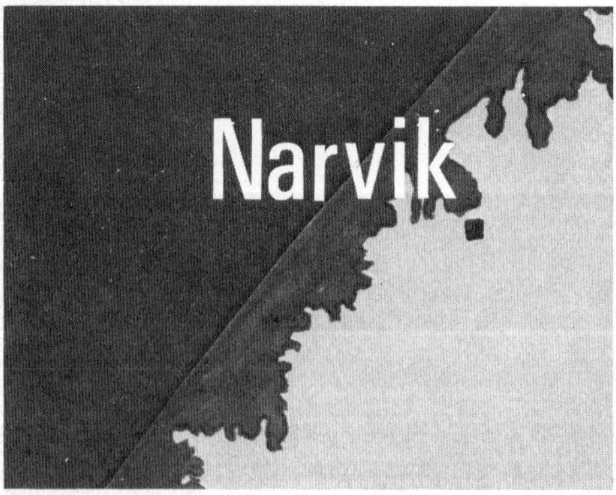

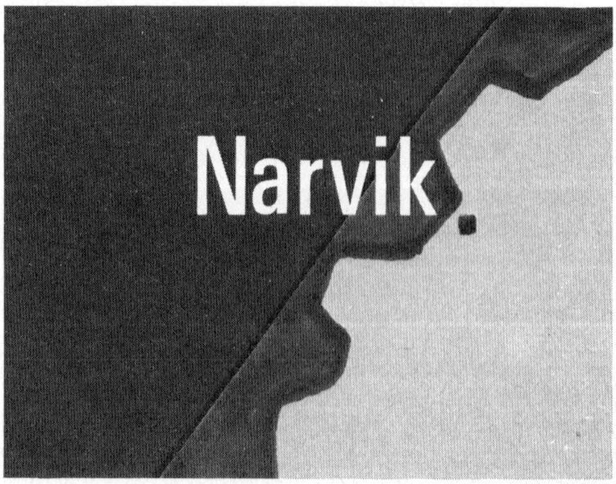

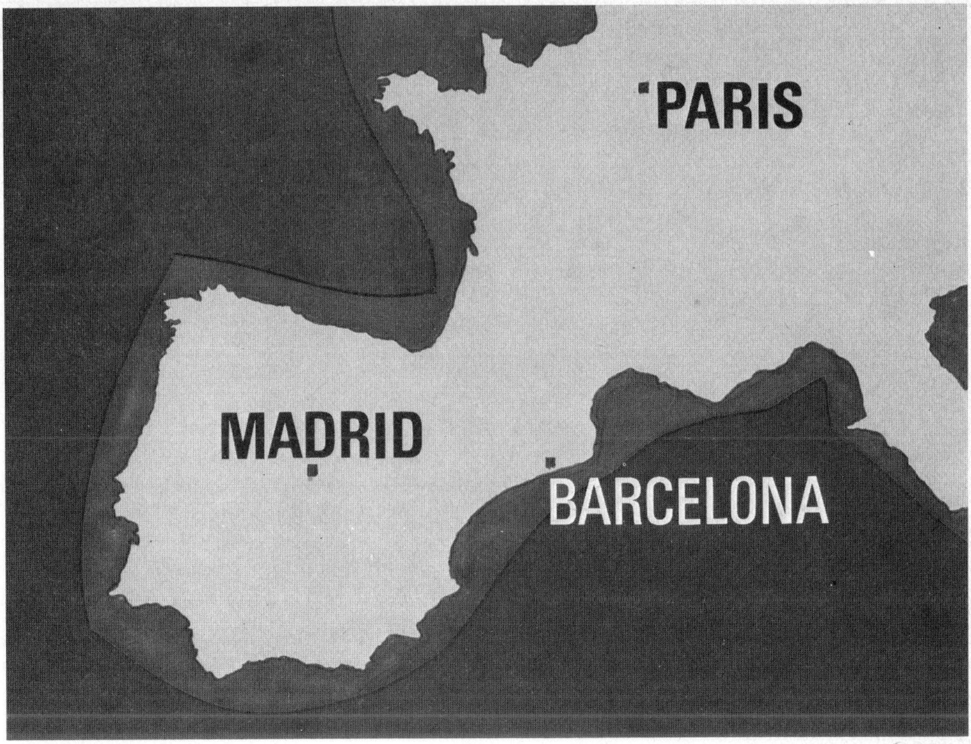

of programmes. One of its most common uses is on maps for national and local news and current affairs programmes. Maps must be clear so that they are easily recognized, and they must be produced very quickly because fresh news may break only minutes before going on air. Therefore, the designer must work out a system, so that he does not have to waste valuable time in thinking about the problem, nor have to wait for something to dry. This type of work must be almost instinctive.

The designer for such programmes must have a comprehensive range of reference, including atlases, ordnance survey maps and other maps of various scales. From these, sets of tracings can be prepared with motorways, major towns and some other features marked, for large or small areas. Some simplification must be made in producing these tracings, because it would take too long to reproduce all the complex features of a particular map. Simplification should be minimal, so that the basic characteristics of the area are not lost, and it can be easily identified by the viewer.

The convention for maps is to have the land masses light or mid grey and the seas and rivers dark grey or black. For colour television the land is orange and the sea dark blue, which reproduces as mid grey and almost black on monochrome television. Sometimes this is reversed when an animation is required on the land mass. The normal method of production is to trace down the map outline on to a grey

The map of part of Europe shows the standard conventions for the production of most maps. The land mass is cut from a sheet of grey card, and stuck to a black 12 in. × 9 in. caption. When cutting the card, do not bother about the coastal indentations, they can be drawn in with a black fibre-tip pen later. This will not show on transmission, because the caption is electronically adjusted, and joins on black disappear.
The lettering is clear and bold (in this case Univers 59 and 67).
Maps are used a lot for news and current affairs programmes, and they need to be produced very quickly, hence the cut-and-stick methods.

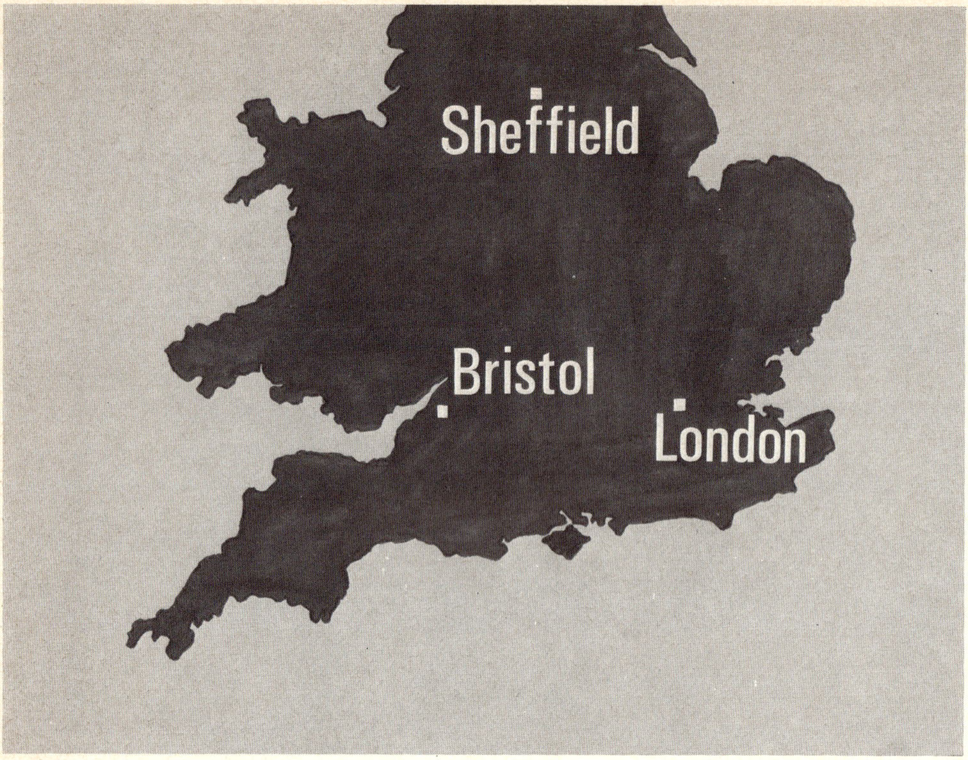

Sometimes the conventions are reversed, and the land mass is dark and the sea light. This is necessary when a name is animated on the map, or when matching a diagram which is predominantly dark.

The maps of Australia (*opposite*) show how names can be emphasized on grey by the use of black or white labels. They are also useful when there is not sufficient room for the name in the correct place: the label can be placed as near as possible and an arrow added to point out the correct location.

card, and then cut it out with a surgical knife. There is no need to cut round every indentation; make fairly straight cuts, since the indentations can be filled in with a black felt-tip pen later. Once cut out, apply Copydex or double-sided Sellotape to the card and fix to a 12 in. × 9 in. black caption. At this stage draw in the rivers, coastal indentations, motorways and railways, etc. The next stage is to Letraset the names of towns, villages and counties. It is best to use only two type styles, such as Univers 57 and Univers 55 or 67, which will give some variation in weight. Black and white letters can be used, although the choice will vary according to whether light or mid grey is used for the land mass. For light grey the normal method is to letter the names in black and where emphasis is needed, letter them in white on a piece of black card, which can be cut out and stuck in the appropriate place. Letters on the sea areas will have to be in white. If emphasis is required when mid-grey land masses have been used, then black lettering on a white label is the convention.

The normal size of lettering for a 12 in. × 9 in. map caption is 48 and 36 point in both upper and lower case; if only a few names are necessary then 48 point would be used, and where more are required a combination of 48 and 36 point.

One of the most important considerations for a map is that it must not be too cluttered with names, but there must be sufficient reference points for the viewer to understand it.

Tennant Creek
Perth
Sydney

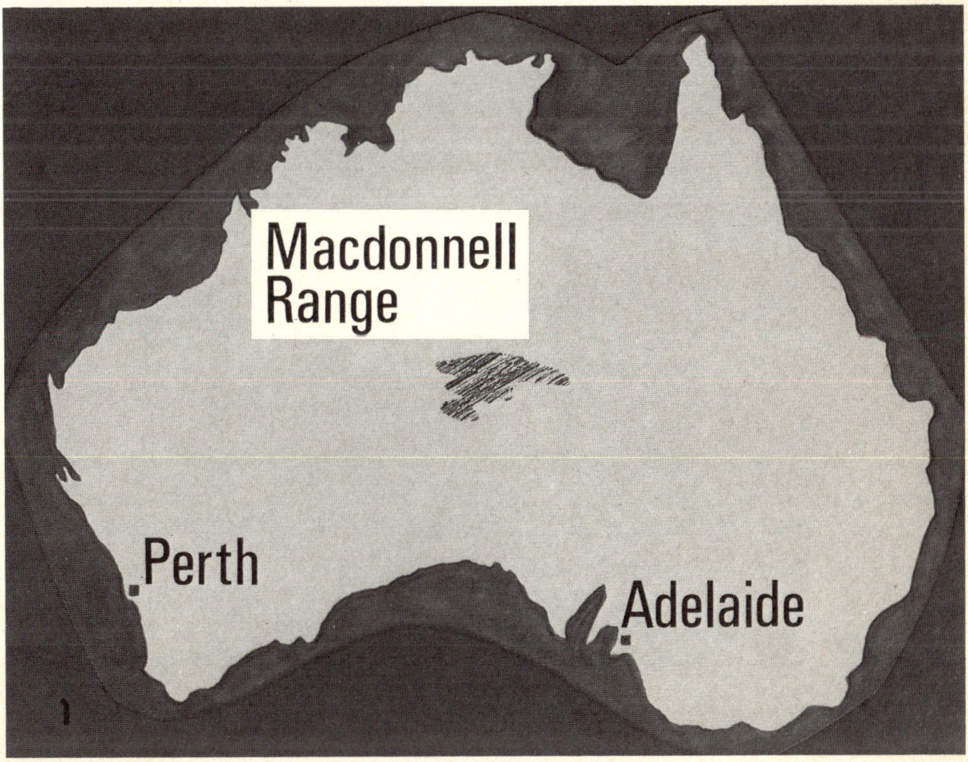
Macdonnell
Range
Perth
Adelaide

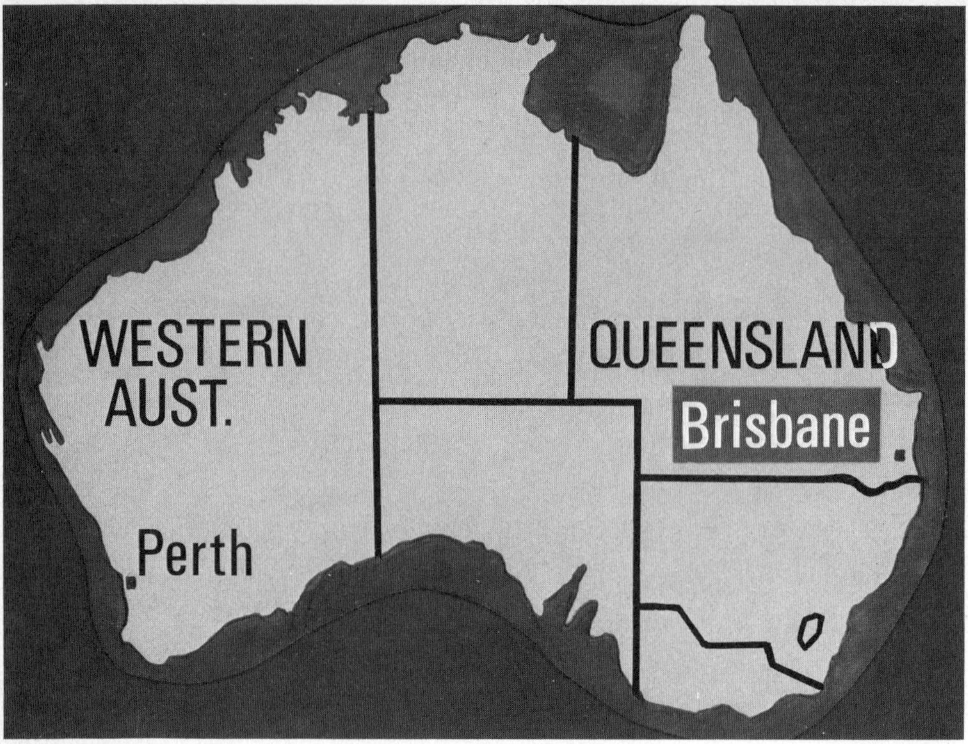

When maps require county, state or national boundaries, the quickest method is to draw a continuous or dotted black line with a fibre-tip pen. If white boundaries are needed, cut white Letraset rules are very suitable.

Usually a large town or city is indicated, so that smaller places can be related to them – people are likely to know roughly where a city is, so they can use this as a reference point.

Some maps need county or country boundaries and the quickest method is to draw them with a fibre-tip pen in black. If white dotted lines are needed, use Letraset white rules, cut at regular intervals before application to the caption. If mountain ranges or specific areas need to be marked, either cut-out tone paper or rub-down textures are the easiest materials to use.

Some maps have information added during transmission; this is achieved by an animated caption or by superimposition. The additional information is produced in white on a black caption, its position being very carefully related to the map. For transmission the two captions must be put on two cameras, and the director then superimposes caption B on caption A, making sure the camera pictures have been matched.

When a correspondent is sending a report from another country by radio telephone, a map is often used to show his location. This map can be livened up by placing a small photograph of the correspondent in a suitable position, which gives an identity to the voice.

Sometimes the scale of a map is such that only land is shown and therefore just a grey caption is needed. In this case it is desirable to use both black and white letters on black and

To establish the identity of a correspondent giving a radio-telephone report, as well as to show his location, a photograph integrated with a map is both informative and attractive.

white labels, to give the map more contrast in its tone range, otherwise it tends to look very flat and uninteresting.

The production of maps tends to be a repetitive job, but there is a tremendous challenge to produce quickly and accurately and to make them as well designed as possible. The author was required on one occasion to produce an outline map of the UK, with two motorways and four cities to be marked, in four minutes, the caption being rushed into the studio with about ten seconds to spare before transmission. What was important about that caption was that the viewer could not have noticed that it had been produced in such a rush, because it was accurate and clear, and did its job in providing information. That incident was unusual, but it highlights how fast the designer may have to think and operate in order to provide one of the visual elements of a programme and not to let the other production staff down.

Two-dimensional diagrams

Two-dimensional diagrams make use of the same materials as maps, but gouache greys and colour paints, textured papers and crayons can be added to the list. Diagrams are often wanted for news and current affairs programmes, but their major use is in educational programmes.

As with maps, clarity is an important consideration, because diagrams are used to explain an idea which is

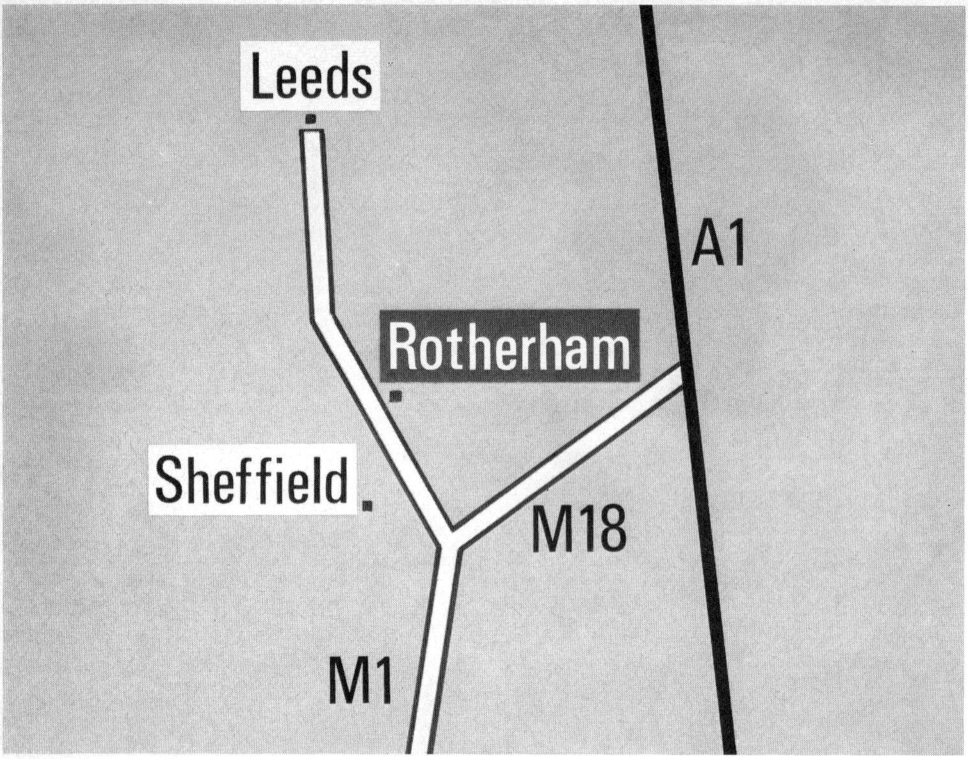

Some maps are only grey land masses, and in these cases it is important to use white and black lettering and labels, to give emphasis to the names and contrast to the caption.

complex or new. Their function is to simplify and explain accurately. There is no point in producing drawings which can only be understood by specialists (for example, engineers), if the viewer is not an expert in the particular subject. This presents the designer with the problem of simplifying the idea without losing sight of the original. To a certain extent, this comes with practice, but it is a good idea to make several rough drawings, simplifying each time, until you feel you have captured the essence of the subject. Quite often the designer will be asked to produce a diagram from a verbal brief only, and there is no time to search for reference material. In these cases he must be as accurate as his visual memory will allow, not particularizing too much in case the details are wrong. A designer must be a walking dictionary of visual information, which he can re-create on paper at a moment's notice. The more drawing and observation the designer can do in his spare time, the easier it is for him to produce drawings from memory. I myself find it useful to keep a collection of notes and sketches on a wide range of subjects, for instant information when working in television.

The examples show a variety of approaches to diagrams; notice that all of them are clear, have a good tone range, give emphasis to certain parts of the subject, and are attractive and well designed. There is also a good relationship between the style of drawing and the subject.

Some diagrams are much better when produced as a white line drawing on black card, particularly if they are simple images. The same diagrams drawn in black on a grey card would lack contrast, and would not be so bold and effective. This type of diagram (white on black) can be superimposed on another picture source, either to add information or to make the picture clearer. Superimposed diagrams must be carefully matched both at the drawing stage and when being televised. Sometimes they are superimposed on film, and this can present some difficulty in obtaining the required information to base the diagram on,

All the diagrams below are variations of the two-dimensional style. They are fairly simple and have a clear tone. The first two diagrams are for biology programmes. The third (*bottom left*) uses cut-out toned paper and accurate pen drawing and the fourth is a photocopy of a map which has been airbrushed to make the river clearer.

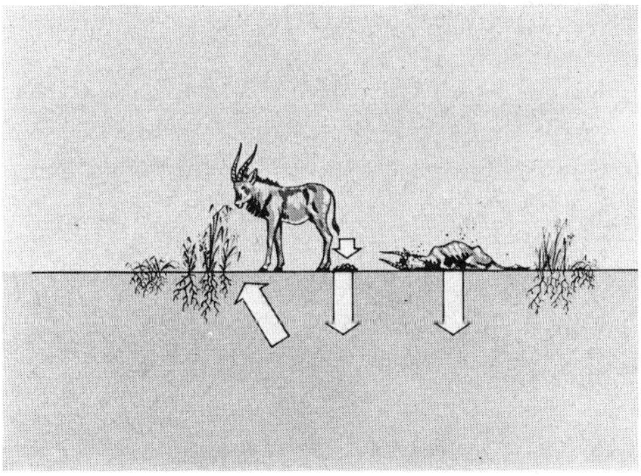

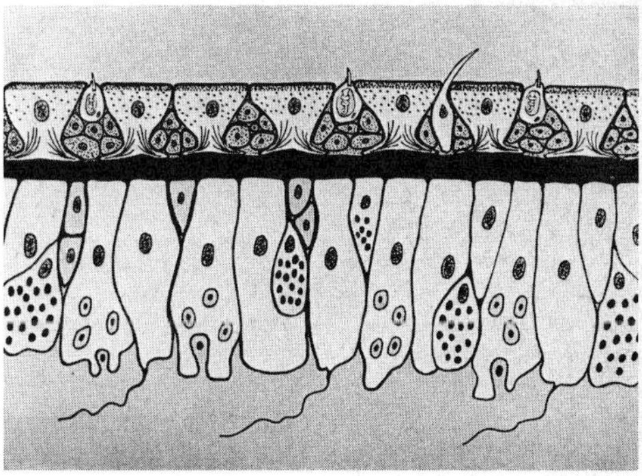

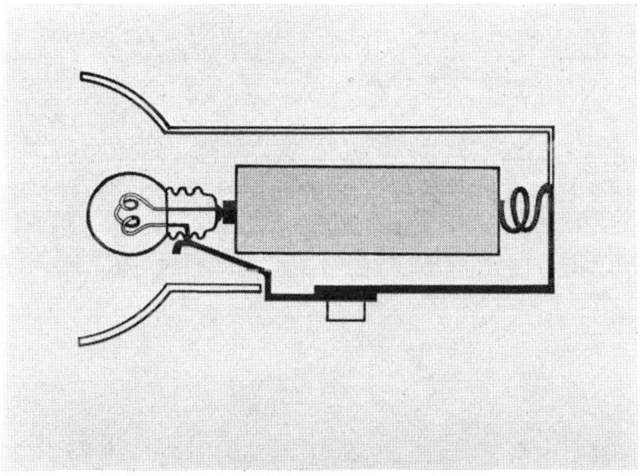

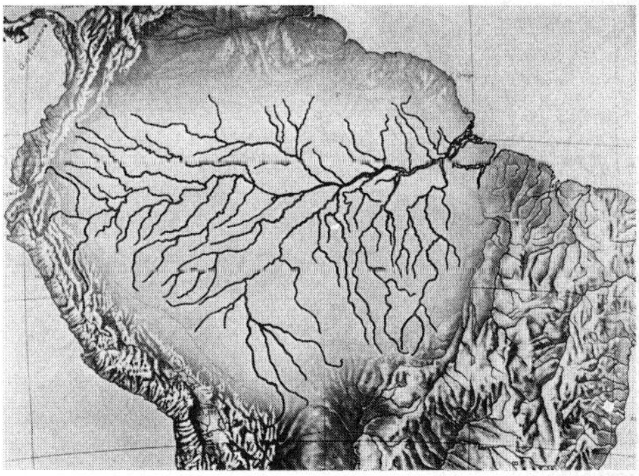

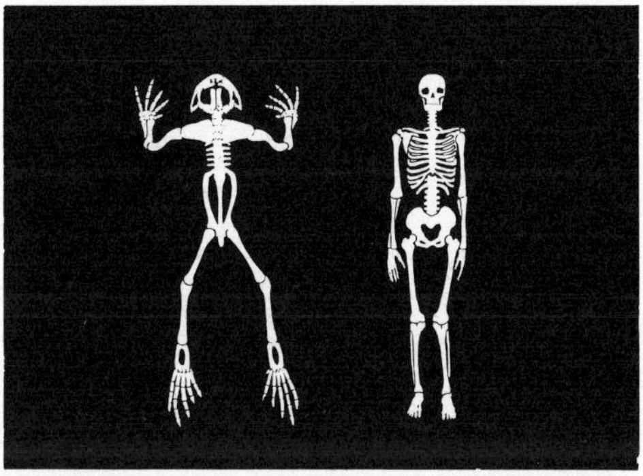

Some 2-D diagrams are much clearer as white on black and they can be superimposed on other picture sources.

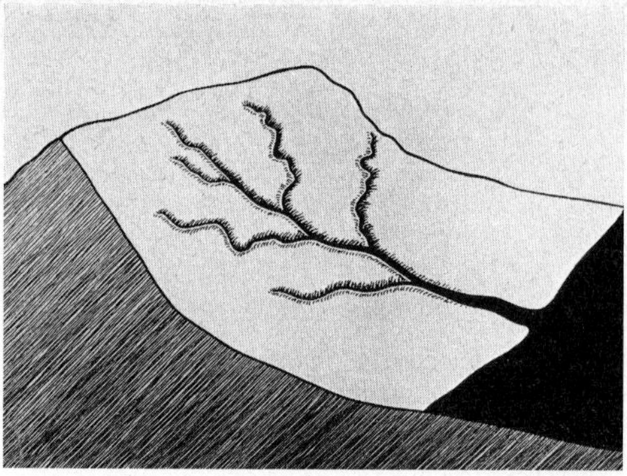

This simple 3-D diagram (*right*) is made from cut-out papers, instant rub-down texture and felt-tip drawing, and clearly shows the course of a river.

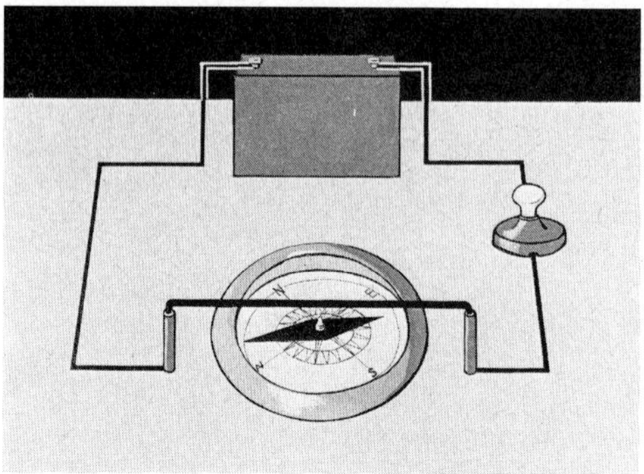

The biology diagram (*above left*) uses a photographic treatment, which requires accurate drawing and a more subtle use of tones. This style of diagram allows the designer to put together elements which could not be photographed easily.
The science diagram (*above right*) uses cut-out papers, paint and pen drawing. Notice the good contrast and arrangement of the elements, which provide maximum clarity. The experiment could have been photographed, but it would have taken much longer to construct than it took to produce the diagram.

because the scale of film and captions is so different. The best way of matching a diagram to a film image is to obtain a clip of the film and put it into a 35-mm. enlarger. Enlarge the image until it is 12 in. × 9 in. on the printing frame, and then trace off the image on a sheet of tracing paper, marking the edge of the frame. The image can then be traced down on to a black card and drawn in white, and should match exactly when transmitted. It is possible to trace off the film image from the screen of an editing machine, except that it will be smaller than 12 in. × 9 in., and therefore will have to be scaled up, either by guesswork or by using a Grant enlarger.

Three-dimensional diagrams

These can present quite a problem to the designer, because the introduction of perspective can make the subject more confused, depending upon the viewpoint or angle chosen to present the subject and the style of drawing. As far as possible the style of drawing needs to be straightforward and simple, without frills or 'interesting' textures which look good but do not add anything in terms of information. The most effective drawings are those that make use of simple tone shapes and some black pen drawing, with perhaps some white highlights. Most three-dimensional diagrams require a degree of accuracy in drawing, and a proficient use of drawing instruments such as ruling pens and compasses. Additional aids are the use of Letraset white and black rules or lines, which are made in various thicknesses and can be applied very quickly.

Apart from the rather stylized diagrams, there is a demand for other styles. One is a naturalistic treatment, which contains a lot of detail and more subtlety in the use of tone. This requires a high standard of drawing and observation, and it has some advantages over the use of photography. It enables the designer to put together elements which could not be photographed together, and which, if they were collaged together, would be unconvincing and messy. The designer has more discrimination than the camera; he can leave out all the unnecessary details and easily control the tonal range and contrast. In some cases his drawings are based on photographs, but the designer should not be bound by the viewpoint of the photograph, particularly if its angle is rather obscure. Captions of this sort take a little longer to produce, to obtain the accuracy, correct perspective and more subtle tone range, but they are a valuable method of presenting information.

An extension of this style is the cut-away technique, which is used to give additional clarity to a subject, and to make a teaching point. This style is very good for showing subjects such as machinery or scientific and biological specimens. Another way of achieving clarity and detail is to have two or three captions of a subject. The first illustration shows the outside configurations, the second has the casing or shell removed to reveal the inside parts, and the third shows further detailed information. The exploded diagram can also be a way of clarifying a complex subject, where the problem is to show the correct relationship of the various parts.

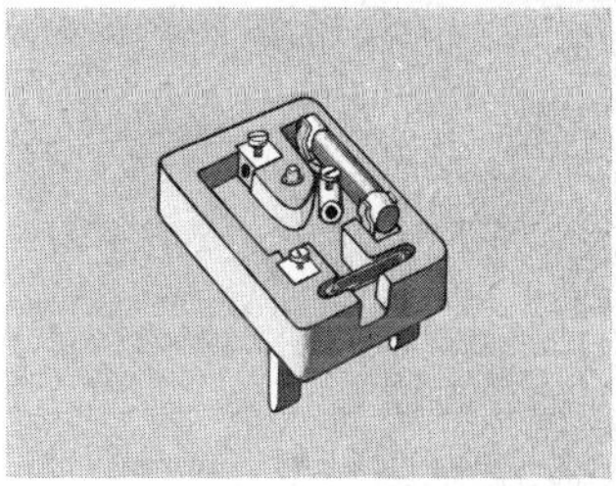

These two captions demonstrate the cut-away technique with accurately matched diagrams. By mixing from one caption to the other, the relationship of the complete object is still retained. The drawing has been simplified a little to give more clarity.

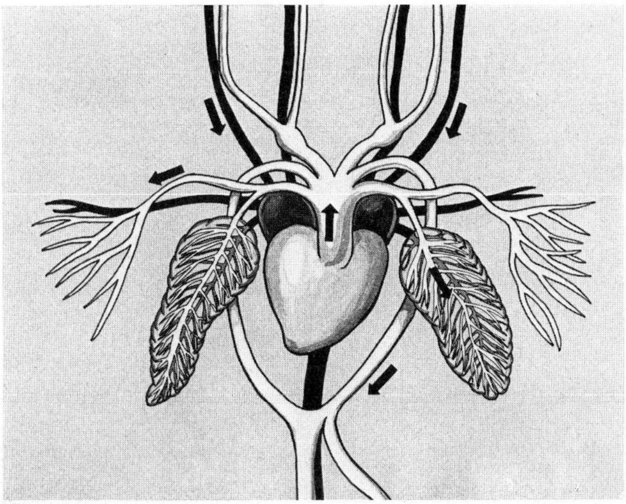

(*Left*) This complex diagram of the heart-lungs circulation system was based on photographic references; photographs could not be used for this caption because they were too confusing. This cut-away technique is very useful for scientific and biological diagrams, but it demands great accuracy and a very clear viewpoint. This diagram was produced with paint and fibre-tip pen.

There is a considerable demand for freehand drawings in a wide variety of styles and materials. Captions such as these are used to illustrate very specific points, to create a mood and interpret themes or abstract ideas. Most types of programme require freehand captions and therefore the designer must be capable of handling a variety of subjects and materials. The drawing of fruit (*top left*) is fairly straightforward, using paints and pen.
The geographical drawing has a much freer style. It was produced with black charcoal and white paint on a grey caption card. Its purpose is to show how a co-ordinate is used to determine the height of a mountain.
The drawing of the soldiers and helicopter was done with a pen and coloured inks, in quite a stylized manner. The coloured inks were used to give some tonal variation, apart from the strength of the line.

With all three-dimensional diagrams, it is important to have accurate reference material, which must then be translated into the best possible terms for television.

Three-dimensional freehand

The scope of three-dimensional freehand drawings and diagrams is almost unlimited, the governing factors being the nature and subject of the programme and the inherent electronic limitations of the medium. This type of work is not limited to single captions; quite long sequences are often needed for illustrating stories or themes, and the captions may therefore be larger than 12 in. × 9 in. to allow panning and zooming.

Such work can range from humorous drawings and cartoons to artists' impressions of historic scenes; from drawings which evoke a mood to very accurate and technical drawings of equipment. It can present the designer with some problems, because for this type of work he must be able to produce drawings almost instantly in a wide range of styles. A producer may want a set of drawings in the style of medieval wood-cuts, or a sequence using comic-strip pop imagery. Whatever the style of the drawing, the main consideration is to be sure that the point to be illustrated is expressed clearly, and that the form of imagery used has meaning, and is the best solution within the given framework of time, reference material and budget. The designer

This caption is designed for the camera to pan and zoom over, and its original size was 24 in. × 20 in. It is made up of some cut-out photographic elements and mainly white drawing. It was used to illustrate a verse of poetry.

The design on the left was for a programme dealing with Iron Curtain countries. The original drawing was 24 in. × 20 in. and produced by coloured wax crayons and inks in a bold and painterly manner. It was one of a series of drawings which were all photographically reduced to 12 in. × 9 in. black and white captions. Colour is often used in producing captions for black and white television, because life would be very dull for the designer if he was always working in greys; but, more important, colour materials offer a wider range of methods and interpretations.

The drawing, for a children's programme, is a simple line drawing on grey, slightly humorous and very quick to produce.

must be sure that his interpretation of the theme or subject is not too obscure or the stylization too extreme. Accepting these considerations, the designer can produce some exciting and creative imagery for almost every type of television programme, particularly children's and educational programmes, but also for promotion and trailer spots between programmes.

All drawings for television demand a bold line, good tonal contrast, clarity of design and lively draughtsmanship. They also demand speed and accuracy. Valuable time must not be spent in experimenting and assessing results if the schedule is tight. Of course, at times the designer does have some time to explore new methods and materials: he must have this time, otherwise his work will become dull and sterile and the process of design mechanical instead of creative. However, a good designer is always looking for new ideas, and in fact these often occur while he is working on other design problems. This is where the designer's memory is useful, enabling him to retrieve information and ideas at a moment's notice, and then to develop the idea. Apart from storing ideas for later use, the designer must observe, analyse, draw and make notes on a wide range of subjects, and information must be absorbed for ideas to be created. He should always be striving to improve the quality and originality of his work, and this is an area where there is no mechanical process which can rival his ability of synthesizing words, thoughts, emotion and information into a visual form. The designer must take full advantage of this ability and push his ideas to the highest possible standard in terms of drawing, style and content.

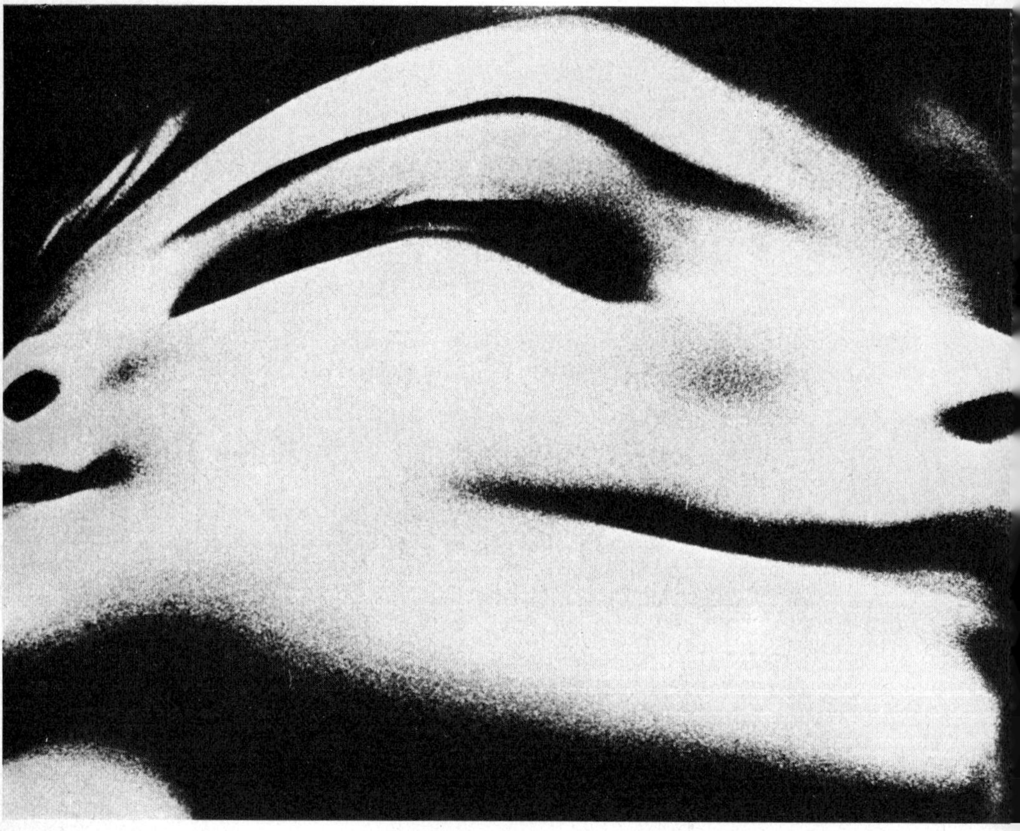

This photographic image was produced by projecting a black-and-white slide on to a plastic mirror (Luminex) which was bent and moved while being rephotographed. The distortion is achieved by reflection from one part of the mirror to another. This type of photograph requires a certain amount of trial and error until the right image is produced.

4 Photography

It is a considerable advantage if the designer is a capable photographer, or at least has some knowledge of the principles of developing and printing and is able to give clear instructions to a technician. Apart from taking and printing photographs, the designer is expected to retouch and manipulate photographic images in a creative manner, particularly for titling and credit sequences and for programme material. Quite often, more than one copy of a caption is required and photography is the obvious means of reproduction.

The designer should accept that photography is a valuable supplement to his range of techniques and creative ability. It is just as important as good draughtsmanship or typography, not only because of the variety of methods and results, but also because of its qualities of communication. Photography used intelligently both in the shooting stage and in the printing can have great impact on the screen. In most cases, its statement is less likely to be impaired by personal styles or whims. There are certain subjects, such as a photo-micrograph of a cellular structure, which can be handled only by photography; a hand-drawn caption cannot hope to match the detail and complexity of such a subject. For programmes such as news, the designer can give only an impression of an incident, whereas the photograph describes it fully.

One distinct advantage that photography has over hand-drawn captions is its ability to integrate smoothly with live action material on film and in the studio. It has the same tone structure as live material, the details are likely to be the same, and where performers are concerned, it can be difficult to distinguish photographs from live pictures (except that they are frozen) as long as the lighting is identical.

EQUIPMENT AND MATERIALS

Ideally, a graphic department should possess a good-quality stills camera and a darkroom with at least the bare necessities – an enlarger (preferably with two lenses), an electronic timer, developing tanks and dishes, a print dryer and of course running water. The camera should be either a single- or a twin-lens reflex, using 35-mm. or $2\frac{1}{4}$-in. square roll

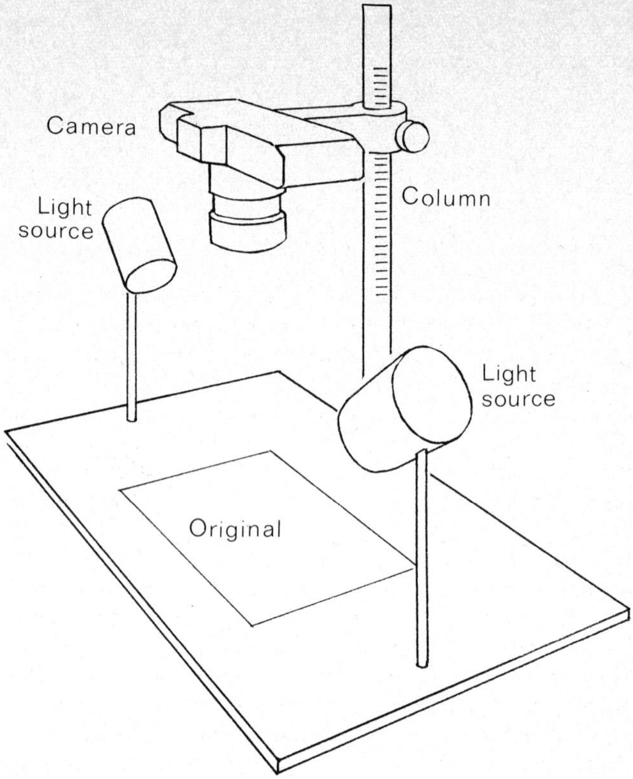

A simple but permanent copying stand with fixed lights is recommended for the photo-copying of captions and other images, and a single- or twin-lens reflex camera. It saves a lot of time, and a set of exposure and development procedures can be adopted to ensure consistent results. Stands like these can be purchased, but they can be easily built by the designer and thus save money towards a better camera.

film, and a telephoto and wide-angle lenses are a distinct advantage. A 5 in. × 4 in. plate camera is extremely useful for one-off shots, and for using a wide range of films. For big enlargements, a 5 in. × 4 in. negative will give better results than a 35-mm. negative. However, a special enlarger is necessary to print the negatives, and these can be rather expensive. Whatever your camera and darkroom equipment, it is essential to set up an efficient operation, so that time is not wasted.

Some of the photographic work will be the copying of material, and a copying stand with an adjustable column (which is calibrated to image sizes) and fixed lights is recommended. This equipment can save a lot of time lining up artwork, cameras and lights, and it enables the designer to work out and observe a set of standards of exposure and development times, for different film stocks, which will ensure consistent results.

It is best to use only one manufacturer's film, paper and chemicals since this enables the designer to work out a set of standards. This does not preclude experimenting with new materials and methods when the subject or programme requires it; in the long run it saves time for the designer to experiment.

Photographs for television should be printed on matt paper, although in an emergency non-glazed lustre or glossy paper can be used. The standard size is normally

12 in. × 10 in., but larger sizes can be used so long as they are in the correct ratio. As with all captions, standardization on size helps to smooth out production problems, so stick to the 12 in. × 10 in. size, except when the photograph is required for panning and zooming. The reason for using matt paper is to avoid the problem of flare from the studio lights, and it also tends to give a slightly flatter tone range, which is ideal for television. Matt prints are much easier to retouch than other photographic print surfaces, and this can be done with pencils, paint or an airbrush. Single-weight paper is normally used because it dries much more quickly than double-weight, and it is less bulky when dry-mounted on card.

The films required are a medium–slow-speed film (25 or 80 ASA) for copying and average lighting conditions, and a fast film (400 ASA) for available light and for dark conditions. These will cope with most work, although a very slow line film (4–10 ASA) is very useful for copying white line on black subjects or for making duplicate negatives. With colour, in order to save time and money, most of the work is done on reversal slide material, although some television organizations do use prints. The problem here is getting the correct colour balance when printing. This can take a lot of trial and error, unless the enlarger has electronic exposure control and there is an automatic processing machine. Unfortunately these are rather expensive and they are economic only when large numbers of prints are produced.

TONE

The tone range of photographs for television is virtually the same as the range for hand-drawn captions, and it must be closely observed. The prints should be far flatter in tone than would be accepted in any other reproductive medium, and generally they should not contain large areas of white or very light tones. Subtle details in dark or shadow areas do not register very well on matt prints, so care has to be taken in lighting a subject, and in the printing stage.

Photographs have to observe the same tonal range recommended for hand-drawn captions. They should have a clear tone range, not be too contrasty, not have large areas of white, nor too many important details in the shadow areas. Matt photographic paper should always be used, to prevent flare from studio lights, which can occur with glossy prints. Matt prints generally look a bit flat in comparison with normal prints, but they are much easier to retouch.
The photograph of the bicycle dynamo and lamp (*below right*), is clear and crisp with a good tone range. The shot of aerial crop-spraying (*below left*) is too contrasty and the details in the shadow area do not register.

This photograph of an accident is a typical example of the pictures used in news coverage, and normally would be supplied by a news agency or free-lance photographer. Quite often these photographs are not the ideal tone range or size, and owing to the speed of production there is not much time to retouch them before transmission. However, a contrasty photograph is better than a reporter talking to the camera.

The tone and colour range of colour photographs is very critical, because the designer must always be aware of the monochrome equivalents. Two colours can be very different, but when converted to black and white, they may be the same tone and may not register as separate shapes. Photographs of subjects which are mainly variations of one colour should be avoided. Most colour slide films have a good built-in contrast and therefore reproduce well in monochrome.

PHOTO-COPIES

A large proportion of the photographs used in television are not shot by the designer, nor are they designed by him. News, current affairs and educational programmes have a large demand for photographs. Most of these are initially produced by photo-journalists, agency photographers and free-lance photographers, many of them being cabled pictures from another part of the world, which are distributed by news agencies. All television companies have extensive stills libraries covering personalities, politicians, places, sports events, etc., and reprints are readily available to the producers and designers. Most of the photos used in news and current affairs presentations require only mounting and perhaps some simple retouching. They are normally needed at such short notice that there is no time for high-quality work. So long as the information is clear they will be used, that is if film or videotape pictures are not available. Sometimes the quality of the prints is not very good owing to the very hurried conditions of instant news, but reality, rather than immaculate prints with exactly the correct tone range, is the need. However, the aim of quality should always come first, if time permits. The designer will also use photographs for split-screen captions which have some lettered information for news programmes.

Photographic material is used in many types of programmes for illustrative purposes, and includes stills of people, places, events, historical occasions, animals and subjects not readily available to the producer or designer. Such subjects might come from the station's own library, photo-agencies, information services, industrial, educational and publishing organizations, as well as from the collections of museums and specialist photographers. Some of these sources can provide 12 in. × 10 in. matt prints; otherwise they will lend glossy prints, which can be copied. It is normally the production assistant's responsibility to acquire the relevant prints, from which the designer would choose those most likely to reproduce well on television. Quite often the material available is not exactly right or it has irrelevant details: it could be copied, printed and re-touched, or the tonal balance could be altered by using a different grade of paper.

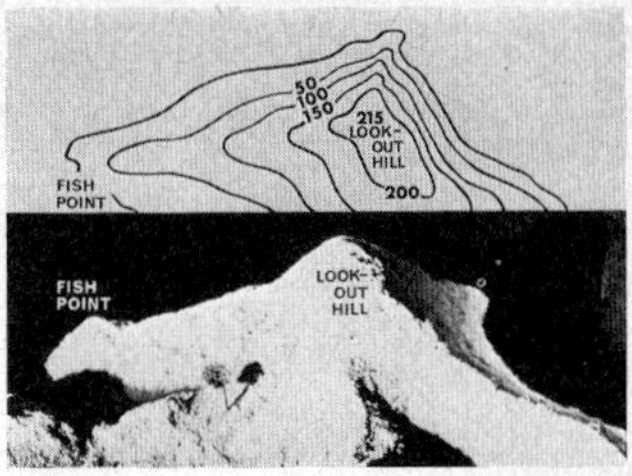

This caption is an integration of a photograph and a simple contour map. This type of caption is often used to help clarify a point in a script.

Riots in revolutionary Petrograd in 1917 is the type of subject not readily available to the producer or designer. It would be found in a photo library, a museum or private collection. Normally the designer is supplied with a print from the original negative, or a photo-copy, but if only the original print is available, photo-copy it (under a sheet of glass to prevent accidental damage), then make the print for the programme, and compare it with the original to check the details. Never use an original like this in the TV studio.

It is often necessary to retouch photographs because the tonal range is wrong and some of the details are bleached out or too dark. A lot of the details are missing on the top photograph, but by retouching with an HB and 2B pencil plus a non-greasy white pencil, many of these details have been restored and the general lightness of the typewriter toned down, as seen in the lower photograph. Retouching needs a little practice, and the golden rule is: be controlled and do not apply too much pencil carbon; it is much easier to put more on than to take it off. After pencil retouching, be sure to apply a fixative, so that it does not smudge.

Retouching

One of the common faults of photo-copies is incorrectness in the tonal range; there may be too much dark or shadow area, or too much detail which conflicts with the main information. Once a matt print has been made, retouching is done by lightening the shadow areas with a non-greasy white pencil. This is applied in gentle strokes; the marks are rubbed with a soft rag, thus blending the retouching with the photograph. Several layers can be built up until there is sufficient detail and lightening of the dark area. In some cases when the area is rather large it is better to use designer's grey gouache, but there must be some gradation of tone, otherwise it will look 'stuck on' and false. This method requires practice so that the retouching or drawing integrates with the photograph and improves the clarity of the picture.

Some photographs have backgrounds which conflict with the subject, and the area may be too large to retouch successfully. This can be overcome by making two identical prints and mounting one on card. Then dry-mount a piece of tracing paper on top. Take the second print, cut out the subject accurately and stick it on the mounted print, exactly matching the position. As can be seen, the background will have lost its dominance and the subject will now be much more important and clear. This method is only successful if the cut-out is not too complicated in outline.

Another method is to dispense with the original background of the photograph, and to mount the cut-out

Some photographs can have a background which is too dominant (*above left*). Make two prints exactly the same size, one dry-mounted and with a sheet of thin tracing paper dry-mounted on top. The main subject, in this case the tower, is cut from the second print, and then stuck on to the first print, carefully matching its position. The photograph above shows the finished result with the background toned down.

Some subjects are photographed against the wrong background, perhaps too close in tone (*left*). If the shape is fairly simple it can be cut out and remounted on a darker background, thus making the subject clearer and the tonal range wider.

By using photographic cut-outs and a collage method of arrangement, the designer can exploit the design possibilities of false backgrounds and exaggerated perspective. In this case the caption was one of a series for a title sequence. It is made from three separate negatives – a shot of a window, of a woman's hand, and of some buildings. The prints of the hand and the buildings had to be cut into rectangles, so that they would fit into the window panes. All the prints were made on matt photographic paper, and when cut and mounted were retouched with a black fibre-tip pen and HB pencil.

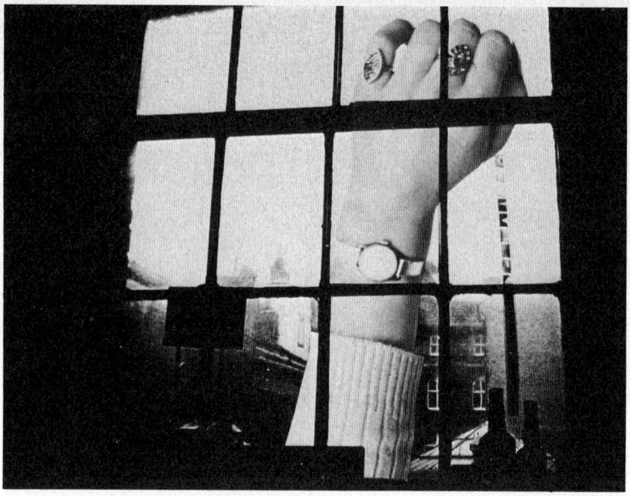

subject on a black or grey card. In other cases, a new photographic background can be substituted, but to achieve realism the angle, perspective and lighting need to be very accurately matched, otherwise it will look false. This falseness can be exploited in some cases such as in titling sequences, and may be extended into collage imagery. Where photo cut-outs are used, all edges must be securely stuck down and all joins retouched with white or black pencils, so that they match the background.

Sometimes light or white areas need darkening to define a shape and to give more contrast to a print. This is best done with black-lead pencils, HB, B and 2B grades, gently applied to the print and smoothed down with a soft rag or finger. Never press hard with the pencils, because this will give a streaky effect rather than a smooth tone. The disadvantage of this method is that the pencil can smear across areas which do not need darkening. A rubber can remove some of this but it is not accurate enough. The best way is to take a thin piece of paper and tear off a strip, then rub a soft pencil on one side to leave a carbon deposit. Place this paper on the print, covering up the area which does not require retouching, with the edge of the strip coinciding with the edge of the shape that needs darkening. Gently rub the pencil deposit off the strip on to the print, moving the strip round to follow the edge of the shape. This will give a soft gradation of tone and a sharp edge. With all pencil retouching you should spray with a fixative on completion, otherwise it will tend to smear. Ideally, it is better to rephotograph and make a new print; then there is no danger of damage.

For darkening light areas in a photograph, rub some pencil carbon on to a strip of paper, and then gently rub the carbon off the strip with a finger on to the photo.
By moving the strip the contours of the subject can be more clearly defined.

It must be remembered that all prints must be checked for white dust spots and hairlines before transmission. These can be removed by using pencils between HB and 3H grades, according to the darkness of the tone.

All the above methods are quite simple and do not require sophisticated equipment. A more complex method of retouching is with an airbrush, a rather delicate instrument which sprays paint on to the print, producing fine-grain tone areas. The airbrush uses designer's grey paint mixed to a liquid consistency. Before spraying it is necessary to mask off the areas which do not need retouching, either with cut-out paper masks or by painting on a special plastic masking solution with a brush. When spraying never hold the airbrush still; move it across the print in a stroking action, gradually building up the required density of tone. Once the paint is dry, the paper or plastic mask can be peeled off. Airbrushing is very good when large areas need lightening or darkening, as it gives very even results. When finished, great care must be taken to avoid finger marking or rubbing the print before copying. Airbrushing can be used with both matt and glossy prints, but its main disadvantage is the cost of the airbrush and the compresser needed to drive it, as well as the preparation time.

ORIGINAL PHOTOGRAPHS

A fair proportion of the designer's work will be the shooting or directing of still photographic sessions, both in the studio and on location. Stills are used for titling and programme inserts and for record purposes (to ensure continuity between location and studio material in drama, etc.).

It is a great advantage if the designer can shoot his own photographs, because he can then control the whole operation from start to finish. In taking photographs, always choose the right film for the job. Do not use fast-speed film in harsh, bright sunlight because the resulting prints will be too contrasty and some of the lighter tones will be bleached out. Even when using medium-speed film, be careful of strong shadows on the main subject, otherwise the print will need retouching. A reflector made of hardboard covered with aluminium foil, or even an electronic flash, can be very useful for reflecting some sunlight into the shadow areas. Unless a silhouette effect is required, never shoot subjects which are too contrasty or directly into the sun.

Shot composition and framing, and also allowing for cut-off, are very important when lining up shots. It is easy to forget this, and when it comes to the printing stage it is impossible to avoid part of the main subject being in cut-off. Always allow plenty of air space round the subject; if it is too much, you can always enlarge the negative a bit more, but you cannot reverse the process. When photographing subjects (particularly people) make sure that the background is not too obtrusive, and that trees are not seen as growing out of people's heads. It should always be possible to find the best angle, and thereby avoid having to retouch the print.

This is an interesting and well-composed photograph, but it is not ideal for television because there are large areas of white and the photograph is too contrasty. Printing it on a softer grade of paper might be a little better, but the best solution would be to shoot this photograph with a light- to medium-tone background.

When shooting original photographs it is an advantage to have both a wide-angle and a telephoto lens. There are many occasions when it is not possible to get near or far away from the subject, and these lenses enable you to achieve the effect. If a very candid type of photograph is required, when the subject must not be aware of the photographer, then a telephoto lens will allow you to work at a distance, and still fill the frame with the subject. It is possible to use a standard lens for the same type of shot, but it would be necessary to enlarge the negative so much that the grain structure would become obvious and the definition poor. The telephoto lens can also be useful when it is felt desirable to have a shallow depth of field, thus putting objects in front of or behind the subject sharply out of focus. This lens tends to flatten normal perspective, particularly when the aperture has been well stopped down (*f*11 to *f*22), which increases the depth of field. Objects which are far apart then appear to be close together. The selectivity of this lens can be used to great advantage.

The wide-angle lens allows you to obtain a distant view when in fact the camera is quite close to the subject. It also has a very large depth of field, thus providing a sharp focus on subjects far apart. The problem with a wide-angle lens is that it tends to distort the subject, either when operating at close distances or when shooting distant subjects. It is particularly noticeable when the shot contains verticals,

(*Right*) A typical wide-angle shot with the verticals all sloping, one of the features of this type of lens; this distortion can be used expressively for some subjects, but in many cases it detracts.

This photograph (*left*) was taken with a standard lens (55 mm.). The photograph below was taken from exactly the same spot, using a telephoto lens (400 mm.). A telephoto lens is very useful for candid photography; the photographer can shoot from a distance and fill the frame with the subject without being observed. Notice how this lens compresses the perspective. This feature is very useful for some subjects.

such as buildings, because they will lean away from the centre of the frame. When working very close to the subject, the nearest part to the camera will become larger and out of proportion to the rest of the subject, which can be disconcerting.

Whatever the subject of the photograph, the choice of lenses should be considered carefully, always remembering that the subject is more important than the technique used. Lenses can be used creatively so long as the finished result is clear in its information. It is easy to be seduced by the qualities of a lens, and fail to see the importance of the subject. When shooting photographs on location, either out of doors or inside a building, it is often difficult to obtain the ideal lighting, because the sun may be in the wrong position or it may be necessary to shoot directly towards a window. This problem can be alleviated by using a portable quartz-iodine hand lamp, powered by a rechargeable battery. The light can be directed on to the shadow areas as a fill-light, and thus give some modelling while retaining the depth of the subject.

For studio work it is an advantage to have a permanent set-up and floor area, with background paper rolls of different colours, stands and tables for experiments and objects, and a range of lighting equipment. Background papers should be mounted on fittings attached to the wall, so that they are out of the way. A translucent screen made of white polythene stretched on a wooden frame can be very useful for diffusing lights to avoid harsh shadows. At least four 500- or 1000-watt lighting units are needed, on adjustable stands and with barn doors. If it is possible, a few extra lights can be suspended from the ceiling on adjustable fittings. These are good for top-lighting subjects and also for high-angle back-rim lighting. A few small 150- or 200-watt lights with barn doors and snoots are very handy for lighting models and experiments, and for extra fill lighting. This type of equipment will cope with most eventualities. A good do-it-yourself item is a cove, which can be constructed from a few pieces of 2 in. × 1 in. timber and hardboard or thin plywood. Paint this light grey or white and it will provide a shadowless background for photographing objects.

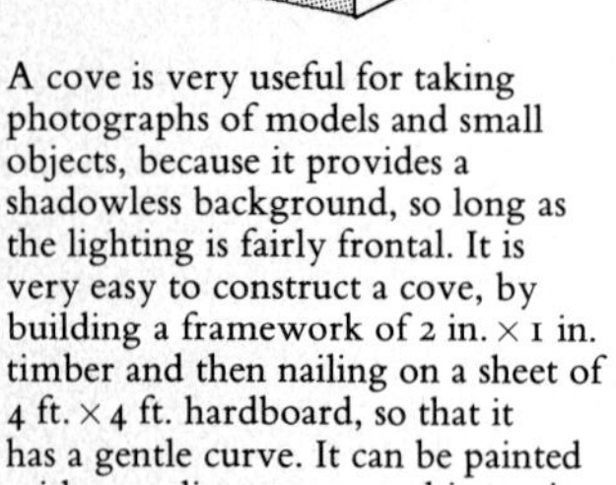

A cove is very useful for taking photographs of models and small objects, because it provides a shadowless background, so long as the lighting is fairly frontal. It is very easy to construct a cove, by building a framework of 2 in. × 1 in. timber and then nailing on a sheet of 4 ft. × 4 ft. hardboard, so that it has a gentle curve. It can be painted with a medium-tone emulsion paint or draped with a piece of background paper.

When lighting the subjects, have your main source of light coming from one angle (key-light) and use additional lights to fill the shadow areas and background. The background lights should never be in front of the subject, otherwise multiple shadows will be created. The further the subject is away from the background the better, because then the key-light will not cause shadows on the background. Shadows can detract from the clarity of the photograph and can be very irritating.

When taking light readings, be sure to measure the light reflected from the subject, and not from the background, otherwise the subject can be under-exposed. When the

These are some record stills of location filming for inserts in a drama production, often vital for checking the continuity of actors' costumes and make-up, so that the film and studio material matches. These photographs can be shot on a 35-mm. or Polaroid camera. The advantage of the latter is that the photograph is developed in the camera and the print can be inspected within sixty seconds. This enables the photographer to reshoot immediately if the first shot is not adequate.

subject is against a black background, and you want it to stay black on the photograph, make sure there is no spillage of light and give the subject absolutely minimum exposure.

As mentioned earlier, record stills are often taken of actors on location, so that there is continuity between the location and studio material. These have to be shot quickly, so that the production is not interrupted, and they must be accurate records of the scene. One method is to use a Polaroid 'instant' camera; the photograph is developed in the camera and can be viewed within sixty seconds. If the shot is not right, it can be reshot at once. This has advantages over the normal methods, but the materials are rather expensive; also the print is only 5 in. × 4 in., and it is not possible to make further prints or enlargements because there is no negative.

Photography is often used to produce insert captions for many types of programmes, and also for sequences and titling. Some programmes use a photo stills sequence as a means of telescoping a particular action, to create a mood which is different from the rest of the programme, or for a dramatic effect. A photo stills sequence is much cheaper than a film sequence, and in some cases it is much more appropriate than film because of its stylization. These stills can be animated by juxtaposition of shots, panning, zooming, mixing and superimposition, either by the television cameras in the studio or by a rostrum camera. This type of

1

2

3

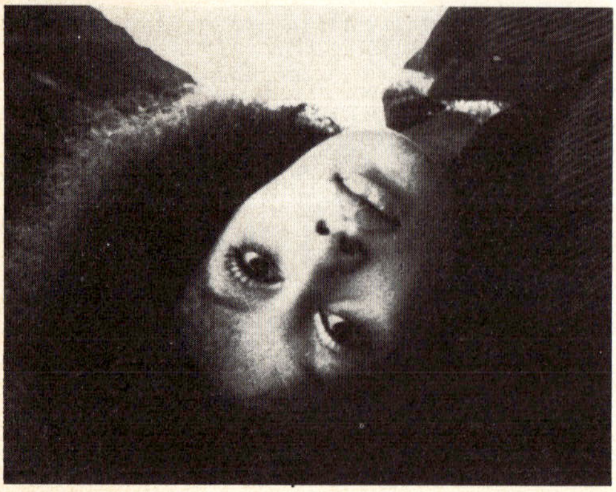
4

5

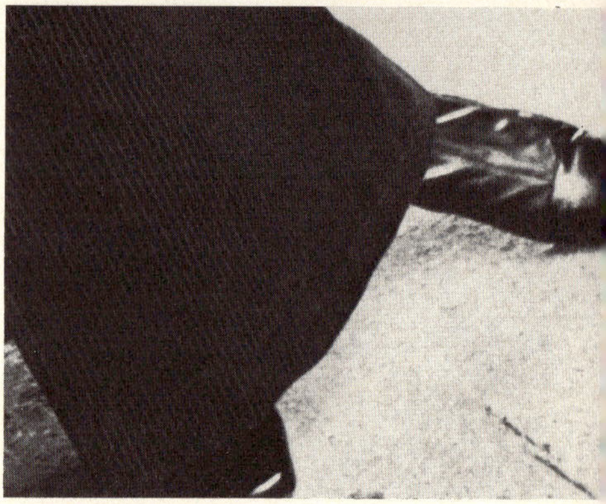
6

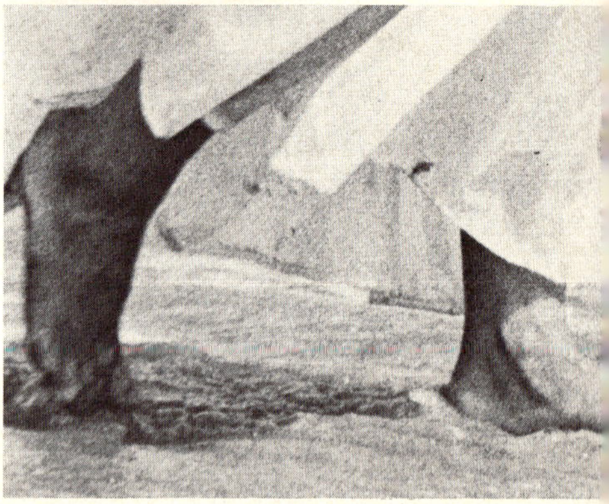
7

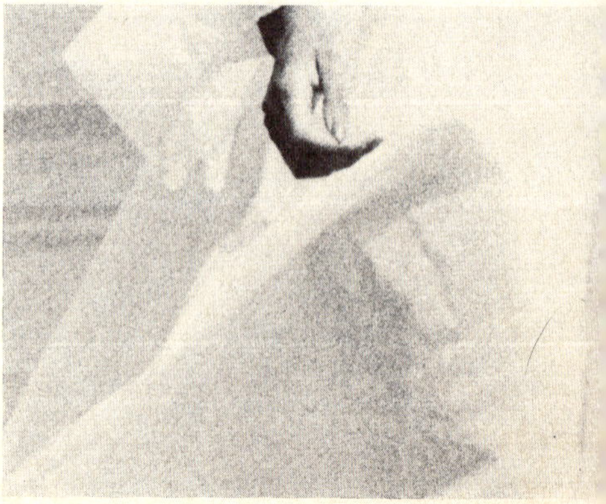
8

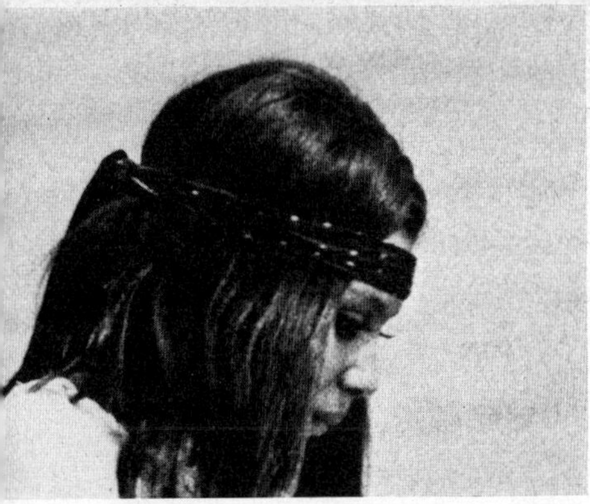

9

12

10

13

11

14

work is often used in drama, light entertainment and educational presentations.

If the animation is done in the studio with television cameras, the stills need to be larger than 12 in. × 9 in., so that the camera can pan and zoom without danger of going off the edge of the still. The usual size is 16 in. × 20 in. or 20 in. × 24 in., which although not exactly a 4:3 ratio allows sufficient scale for camera movements. With the stills arranged on two caption stands, continuous movement can be achieved by cutting or mixing from each camera, with a caption change when each camera is not on air.

An animated photographic sequence.

(See overleaf)

(Continued from p. 63)
The sequence overleaf was produced with these five stills. By careful framing, flowing pans, zooms and mixes, a dream-like floating sequence of two minutes was shot, for use in a drama. The sequence could have been shot by live-action film, but it would not have had the same quality and the cost would have been much higher. Using animated stills is economical, and it presents the designer with many possibilities for creative ideas.

Photography used for titling and promotion sequences makes use of a variety of methods and techniques. Some of these are fairly normal photographs and the lettering is superimposed electronically when the programme is recorded or transmitted; others have the lettering integrated at the printing stage, or are Letrasetted after printing and mounting.

Whatever the way of putting the lettering on the caption, the photograph must be designed so that the lettering has the right relationship with the subject of the photograph. It must not look 'stuck on', it must not be too cramped and it must not cut across the main information of the photograph.

(*Above*) This chart shows the approximate monochrome equivalents of some of the colours used in graphics for colour television. It is important that the designer should know the monochrome tone values, because although the programme is transmitted in colour, a large proportion of the viewers have only black-and-white TV sets. Therefore the designer must be sure his colour designs have the correct tonal range for monochrome.

(*Right*) With colour maps it is not possible to animate additional information on the caption, because the cut-out edges would show. The method used is to make a basic map caption (*top*), and then produce a separate caption with white information on black card (*centre*). This must be very carefully matched for positioning of the words etc. to be animated. On a cue the second caption is inlayed on the map, so that only the white information registers on the map (*bottom*). This method can be used for a wide range of graphics.

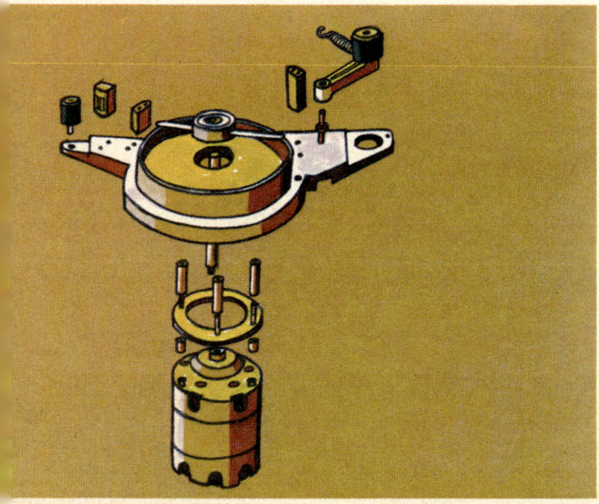

The exploded diagram (*left*) is a very good way of showing the relationship of the parts of a piece of equipment. The viewpoint and perspective must be worked out carefully, so as not to confuse the viewer.

(*Above*) Hand tinting black-and-white glossy photographs with coloured felt-tip pens and inks is an economic and artistically challenging method of producing colour photographs. This method is particularly effective with collage-type imagery for titling sequences and captions.

This series of captions is an interpretation of the Bayeux tapestry for a historical programme. They are simple and evocative, and they present historically accurate information.

Instead of applying Letraset to a photo caption after printing, the lettering can be integrated at the printing stage. This is done by putting black Letraset on cel, which is laid on top of the photographic paper on the baseboard of the enlarger, held flat with a sheet of glass. A negative is placed in the enlarger and about three seconds added to the exposure time. The black lettering is like a negative, and the result is crisp, pure white letters on the photograph.

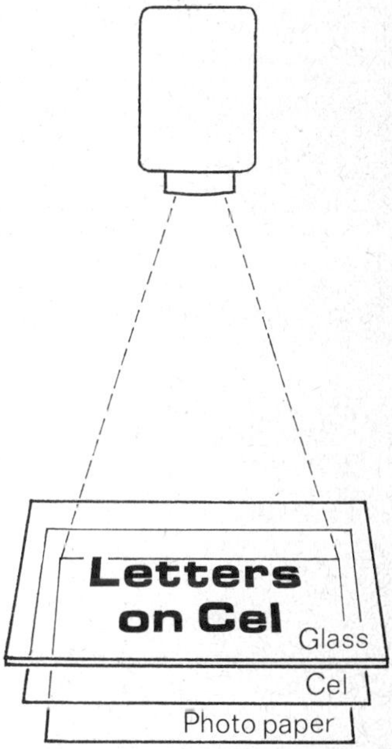

The method of integrating lettering with photographs at the printing stage is an extension of the production of white letters on a black background. Put black Letraset on cel and then contact-print on photographic paper. A negative is placed in the enlarger and framed up, photo paper is placed on the masking frame, then the cel with black Letraset is laid on top and held down by a sheet of glass. The photo paper is then exposed normally, except that two or three seconds need to be added to the exposure time. The resulting print will be a normal photograph with pure white lettering, which is very crisp. It is always necessary to position the lettering in an area of the photograph which is fairly dark in tone, otherwise there will not be enough contrast for the lettering to register. This method gives better results than by applying Letraset to finished photographs, because the white letters are really opaque and there is no danger of the lettering being scratched.

Apart from straightforward photographs, there is a demand for a more creative use of photography, which can be achieved in the development of the negatives or in the printing and enlarging stage. This involves mainly darkroom work and the use of different types of films, papers and chemicals. Some of the work needs time for experiment, so that a set of procedures can be worked out and noted for later use.

In the development stage, it is possible to use developers which will increase the contrast or the grain structure of the film. This is useful when the film has to be shot in poor lighting conditions and gives the negative more contrast and better prints. Such developers can be used to increase the contrast of normally exposed negatives and will tend to bleach out the light tones. Some other developers, with modified development times and temperatures, will considerably increase the graininess of the film, particularly when printed on a hard-grade paper.

This print is the result of reticulation of the negative, which causes the emulsion grain to form large clusters. Techniques such as this are always done on copy negatives, because there is a danger of ruining the negative during development.

Another technique is the reticulation of the negative, where the grain forms large clusters, instead of being even. This is done by developing the negative at a higher temperature than normal, then washing the film in very cold water and fixing in a warm fixer solution without a hardener. If the negative is washed in warm and cold water alternately after fixing, this will complete the reticulation. If reticulated negatives are printed on hard paper the effect is emphasized. Reticulation usually involves the original camera negative, but it can also be done using a duplicate negative, thus preserving the original in case normal prints are required.

To make duplicate negatives, a duplicate positive print on fine-grain positive film should be produced by contact-printing the original negative. Once this has been processed (ordinary paper developer can be used), it can then be printed on to a variety of film stocks to make the new negative. If you want to increase the grain structure of a shot, enlarge a small portion of the positive on to line or Kodalith film. This will also drastically increase the contrast of a negative. Kodalith Autoscreen film has an in-built half-tone screen (as seen in newspaper photographs), which, when enlarged, gives a print with exaggerated half-tone dots.

Another method is to print on to fine-grain positive film, to expose the film to normal tungsten light for a few seconds during development, and then to continue deve-

lopment. This will give a solarized effect, reducing the photograph to almost a line drawing. There are variations in the form of solarization, depending on the quality of the original negative. If a dream-like quality is needed for a print, then this method can be very useful.

Apart from making new negatives, further techniques can be used in the printing stage. One is to cut a series of masks from black paper, which will prevent exposure of certain areas of print, so that they remain white. These areas can have another image printed on them if a new negative is placed in the enlarger, and the already exposed areas masked off. In some cases an overprinting of the images is possible. Or it is possible to sandwich two negatives together and then print from them. The negatives have to be very carefully arranged, because a black area on one negative will prevent a white area on the other negative from printing through.

Some negatives may have an area which is a little over-exposed, but it may be possible to print it up by 'dodging'. This involves using your hand as a light mask between the enlarger lens and the paper, and masking off the areas of the photograph which do not need more exposure, while allowing the required part to receive more exposure. The hand should be moved gently so that there is not a hard edge where the print has received more exposure. It is also possible to burn in small details such as faces.

Like the previous example, this print was produced by making a copy negative, except that the negative film had an in-built half-tone screen (Kodalith Autoscreen). This gives an exaggerated half-tone dot when the negative is enlarged. This process tends to increase the contrast of the original negative.

The photograph above was taken in an abattoir, and is a straightforward print. The print below was derived from the same negative, and is the result of solarization of a copy negative. This is achieved by making a positive print on fine-grain positive film, and after processing, printing on to line film. When this is developed, the film is exposed to tungsten light for a few seconds at the half-way stage, then normal development and fixing continued. The print has the quality of a pen drawing.

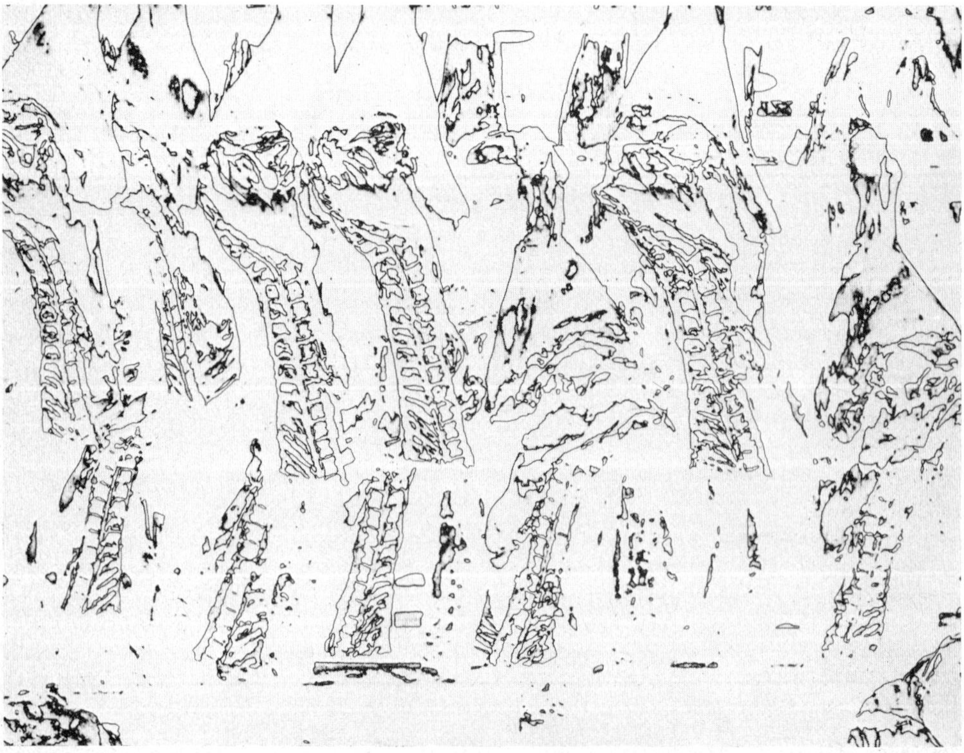

On occasions a series of old stills which are photocopies of the originals may be used for a colour presentation. To improve their quality and to emphasize that they are old, it is possible to make them sepia-toned. A normal black-and-white print is made, and after developing and fixing, it is placed into a bleach solution and gently agitated. The print will gradually become pale, losing all its dark areas. It is then washed and placed in a sepia toner solution and the print slowly regains its contrast, except that it is now a sepia brown. After a brief wash, it should be fixed again and given a final washing. The bleacher and toner solutions can be purchased ready mixed or in powder form.

When colour photographs are required for titling captions and sequences, an artistically good and economic method is to make black-and-white prints, and then hand-colour them with photo tints. These are made in a variety of colours and can be used full strength or diluted with water and applied with a brush. The colour is generally rather stylized, but it is possible to be much more delicate and realistic with a little practice. Where collage-type imagery is used, hand tinting is an appropriate method of introducing colour.

When producing photographs for film animation, it is generally better to print on glossy paper, so that you get a fuller tone range and slightly more detail. Glossy prints present no problems when filming with an animation camera, because drawings are normally filmed on cel, which is just as shiny. When printing stills for filming, be sure to allow plenty of cut-off area. Some cut-off occurs while filming and some more when film is screened through a telecine system, and so there is more than if the photos were on caption stands in the studio.

SLIDES

Instead of having 12 in. × 9 in. photo captions, it is often an advantage to use 35-mm. or $2\frac{1}{4}$-in. square slides, black and white or colour. The slides are shown on a telejector (a slide projector mated to a TV camera), which frees the studio cameras and allows more flexibility in production. It is particularly useful for name supers in an interview, when all the cameras can take shots of the participants, rather than one being locked on to captions. In some cases it is possible to reverse-phase the telejector camera, thus changing a negative to a positive. This can save a lot of time in producing slides, because a normal line negative of the artwork can be mounted and transmitted in a very short time.

There are two ways of producing slides for television: first, the copying of artwork and photo prints, and second, the shooting of slides in the design studio and on location. The second method is rather inflexible, as the shots have to be very accurately lined up when shooting on reversal

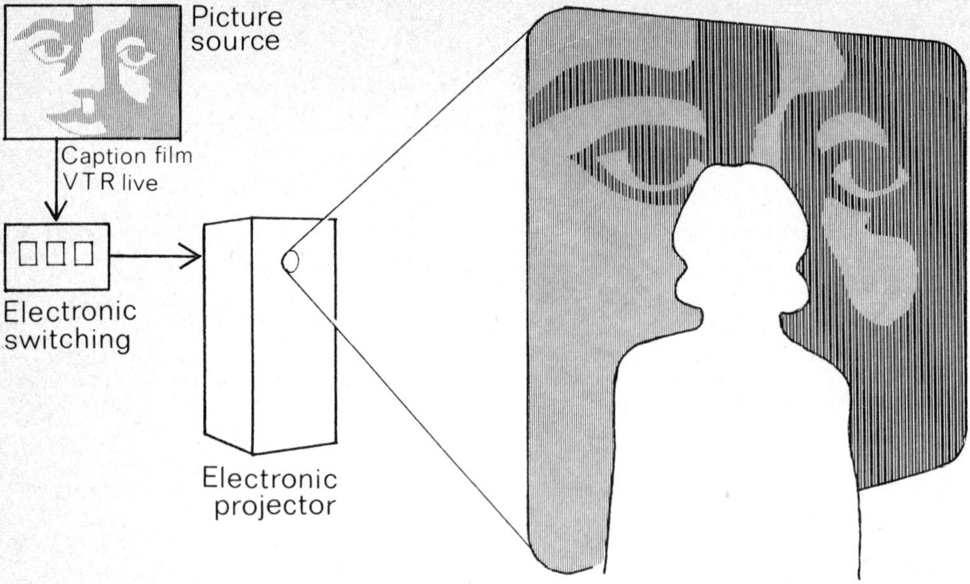

Studio sets often make use of back-projection systems, which use either slides or, in more sophisticated forms, captions and live images. The latter system, called Eidiphor, projects a television image, and is very expensive. The simpler systems are based on a 35-mm. or $2^1/_4$-in. square slide projector, and usually have a blimp to reduce the noise of the cooling blower and change mechanism. Apart from presenting decorative and news information, back-projection can be integrated into sets for drama, etc. to produce views through windows, and travelling backgrounds behind actors sitting in a car in the studio.

material and they cannot be easily retouched, while the first method allows for everything to be exactly correct before the slide is made.

Two types of material can be used for making slides. One is a specially produced reversal film, which can be processed by the manufacturer or in your own darkroom; the other is a line or medium-speed negative film for printing either on glass slides or on to fine-grain positive film. This enables you to have prints as well as slides without any further darkroom work.

Slides are also used in television for front and back projection on screens incorporated into the set. Generally a shot is set up which has a presenter or interviewer sitting beside a screen, on to which a series of slides are projected to reinforce the words of the script. This kind of arrangement is often used for news, current affairs and regional magazine programmes, and the slides have either to be produced very quickly, or to be selected from an extensive library of slides. The Polaroid camera can produce slides as well as prints in sixty seconds, but again the materials are rather expensive.

Another method for front or back projection is to use an overhead projector. This projects images from transparencies approximately 10 in. square. These are made by printing normal negatives on to large-sheet film or with a special photo-copying machine, which uses an original of the same size and prints it on to transparent film. This is mounted on to a cardboard mount, and the proportions are changed by adding black masks. The transparencies produced by this machine can also be used in the preparation of artwork for animation, for titling and programme inserts. They can copy most types of originals, although it is preferable to have fairly contrasty subjects.

Front-projection can be used for presenting information on a screen integrated with the set. For this an overhead projector with a wide-angle head can be adequate. The transparencies can be produced photographically (the originals must be 10 in. × 10 in.) very quickly by machines such as the 3M transparency/photo-copy machine. They can be hand-made by using overlays of rub-down colour and letters which are marketed specially for overhead projector transparencies, or they can be a combination of both methods. The frame format can be altered by masking with black paper, either on the transparency or on the light platen.

LARGE BLOW-UPS

Large photographic blow-ups are often required for set dressing and in the construction of sets, and may be used to create more realism. The normal minimum size for this work is 3 ft. × 4 ft. and probably two blow-ups would be needed for one set. Enlargements of this size can present problems if the darkroom is small, and the enlarger is for 35-mm. films only. To obtain successful blow-ups, the darkroom needs to be over 10 ft. long, and equipped with a 5 in. × 4 in. enlarger. The best method is to mount the column and head of the enlarger horizontally on a trolley, and then project the negative across the darkroom on to a wall. By moving the trolley towards or away from the wall, it is possible to obtain sufficient enlargement. Ideally, the wall should be covered with a large sheet of softboard and painted white, so that the negative can be easily focused. The printing paper, which is made in rolls of 48 in. or 54 in. width, is pinned to the board for the exposure. For developing and fixing, long trough-like dishes are recommended, and the print is rolled through the solutions. An alternative is to apply the chemicals with a sponge, but this tends to be rather messy and wasteful, and the print can be streaky if you are not careful. The washing can be done in a sink if your print washer is not large enough. After thorough washing, drain off as much water as possible from the print, and then tape it to a wall to dry. This will prevent the print curling and shrinking too much before mounting.

The best way of mounting large prints is to cut a piece of $\frac{1}{4}$-in. or $\frac{3}{8}$-in. plywood to the correct size and stick the print down with a cellulose wallpaper adhesive. Apply the adhesive to the back of the print first and then coat the plywood. This enables the adhesive to soak into the print and make it slightly damp and pliable. Now lay the print on to

Photographic blow-ups are often required to integrate in sets, to give information and as decoration. The size of these can range from 24 in. × 20 in. to 96 in. × 54 in. In many cases the designer will not have the facilities or time to produce the blow-ups himself, and will send them to a specialist firm. It is recommended to make a small print with the same proportions as the intended blow-up, and with the desired tone range, and send it with the negative to the firm.
Blow-ups are particularly effective when hung in space, as in these examples, and when the background is fairly dark, with not too strong lighting on the photographs.
Ideally, the presenter should not be too close to the blow-ups, so as not to cast shadows, and there should be some back top lighting on the head of the presenter, which will separate him from the background.
The photographs are hung by nylon thread from the light gantry, and in more sophisticated studios can be lifted or lowered on a boom while the programme is in progress.

the plywood, line up one edge to make it square (this is critical if there is any lettering on the print), and gently smooth down with a soft rag, removing all the air bubbles. The edges of the print should be stretched round to the back of the plywood and Sellotaped. As the print dries it shrinks slightly, enough to make it really flat.

When printing blow-ups for use in sets, do not make them too contrasty and light, otherwise they will distract attention from the actors or presenter. Also, when choosing or shooting the negative, make sure that the shot is not too complicated; simple and clear shots are the best.

Blow-ups can be used to create a scene through a window, so long as the perspective angle is correct. They can be attached directly to the window frame or set back several feet from the window. When lighting the scene you must be careful not to cast shadows on to the blow-up, otherwise the illusion is destroyed. If the blow-up is behind the window, it must be lit separately, but care should be taken not to make it too bright. Blow-ups even larger than 4 ft. × 3 ft.

are used to build sets, but are normally printed by a specialist firm. They are printed in several pieces which require very accurate matching and processing.

It is best to use single-weight matt or mural paper, which is easy to handle and does not flare under studio lights. The prints should be checked before transmission for dust marks and hairlines, which can be retouched with a soft lead pencil or plastic emulsion paint.

To sum up, the more the designer knows about photography, the better he can cope with all the problems of designing for television. Photography can be a very important part of the designer's range of techniques, and if he can do most of it himself, then he can make it a subconscious part of his design-process. The ability to visualize ideas in photographic terms, and to make use of photographic elements, will save the designer a lot of time, both in sorting out the idea and in its execution. Even if he does not take the photographs himself, knowledge of photography will enable him to give clear and direct instructions to the photographer and darkroom technicians.

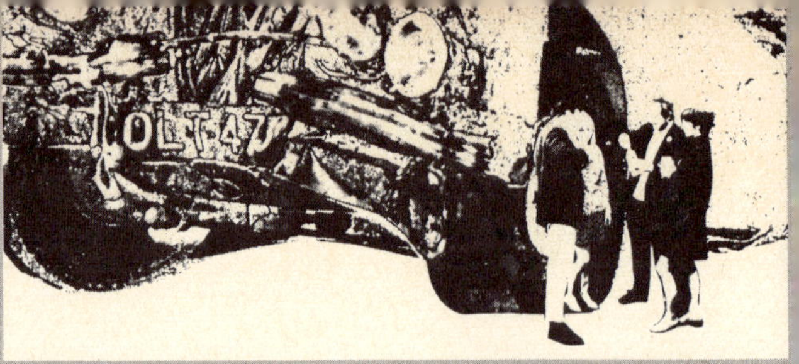

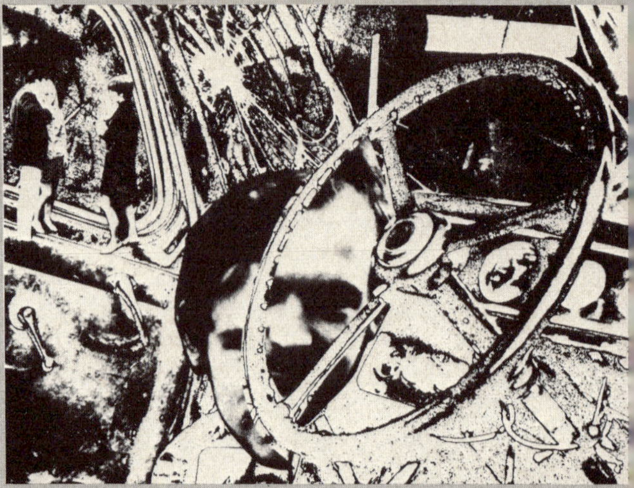

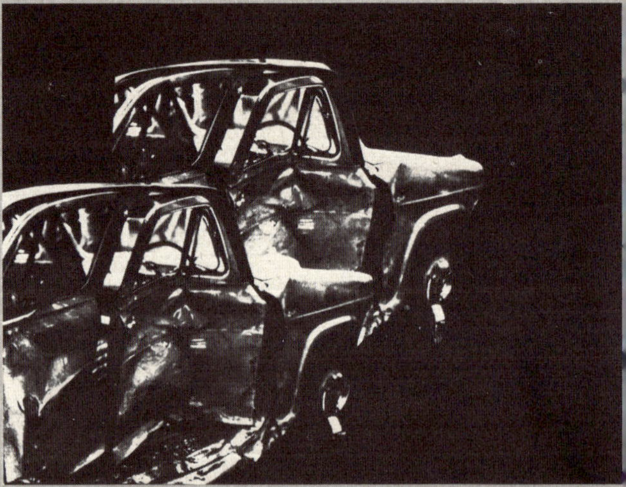

This series of stills is taken from an animated titling sequence shot on film and involving quite a lot of movement of the elements. All the imagery was specially shot still photographs, some made into solarized prints, and the others printed with high contrast. The sequence makes considerable use of cut-outs stuck on cel and exaggerated scale, and was shot to synchronize with some music. It is very difficult to illustrate in static pictures a sequence which involves movement, time and music; it is only possible to show some of the key points and outline the methods, motivation and sources of imagery.

5 Programme titling

Every programme needs titling and credits, whose basis is always some form of lettering, normally added to some other visual statement. This can range from a very simple drawn or photographic image which is on the screen for ten or fifteen seconds to a complex animated or live-action sequence lasting up to two minutes. Whatever the length, style or method, the creation of titling graphics is a vital part of the designer's work. According to the type and subject of the programme, titling sequences give the designer tremendous scope for the development of some creative and exciting imagery. His main difficulty is lack of opportunity, for example when the producer does not want a sequence, or when the programme budget cannot afford a titling sequence. However, the designer can and always should strive for the best possible designs, whatever the limitations or budget, and produce work which is sympathetic to the programme and distinctive in terms of ideas.

Whatever the programme, the designer must familiarize himself fully with the subject by reading the script carefully and discussing the form of interpretation (particularly if it is a drama presentation) with the producer. He must then devise an explicit brief, obtain all the necessary background information and quickly develop his ideas for the programme. There is never much time for the designer to think out the solution to a particular problem, and it is likely that he has work in hand for several other productions. Therefore he has to work out a schedule, which must be very strictly adhered to, if the work is to be completed in time for the rehearsals and transmission. He must, for instance, take into account that some title sequences are to be filmed by stop-frame animation, so that he must spend several days producing the artwork and allow a couple of days for filming; then the sequence has to be processed, viewed, edited (maybe synchronized with music or sound effects) and finally a print has to be made by the laboratories. All these procedures can take from a week to ten days, and that does not allow for the designing time. Whatever the designer produces must be right first time; there cannot be second thoughts because the schedule would be broken and the budget exceeded.

In some respects, programme titling is like the wrapping paper round a package; it can be exciting or infuriating, and the dividing line between the two is extremely fine. The purpose of the titling must always be in the designer's mind; he must never forget that it is a visual statement which must inform the viewer about what he is to see. Clarity and information are much more important than cleverness in design. The range of materials, methods and techniques the designer has at his disposal is very wide. It is sometimes difficult not to be over-enthusiastic and adventurous, and produce a sequence which makes the following programme seem rather flat by comparison. This requires from the designer the indefinable quality of taste, knowledge of what to use with what, so that the sequence truly reflects the aim and level of the programme.

As with most creative activities, the difficulty is not in the method of production, but in the actual design and development of an image or images which are more important than how they are produced. Unfortunately, there is no simple ABC for creative thinking which can be learnt in order to become a good designer. All you can learn, so that they become unconscious means to an end, are the methods of production and the use of many different types of materials and media. The good designer has a repertoire of approaches which he knows will work and which are compatible with his ideas.

The actual design process does not necessarily have any logical path; some of the initial stages can be very irrational as you think around the subject. One designer will quickly decide what the finished design will be, whereas another takes a few components or images and moves them round, constantly modifying them until the design is finished. Both ways are valid so long as you are sure that you have researched the subject fully, rejected all the cliché references, and got below the surface of the subject. In the process you will learn a lot about the subject, which can perhaps be indirectly used in solving another problem.

Every designer has his own method of working out ideas – maybe a few pencil roughs on a layout pad, a collection of cuttings from magazines, or a few photographs. The method does not matter so long as he can sort out his idea clearly and quickly, and explain it to other people.

As an example of the way the designer develops his idea from the brief, the following description outlines one way of developing a sequence.

Subject: ARE LEVEL-CROSSINGS SAFE?

First thoughts: There have been quite a lot of deaths on level-crossings. Have seen reports in the newspapers regularly. Nearly always ones with automatic barriers. Automatic barriers – single poles which swing down just before the train crosses – flashing amber lights – ringing bells. Trains – diesels – engines – carriages – speed. Death – people injured – blood – crumpled bodies – ambulances – gravestones – crosses – crossing – level crosses/crossing. Road – cars – lorries – crossing railway lines – trains crossing roads.

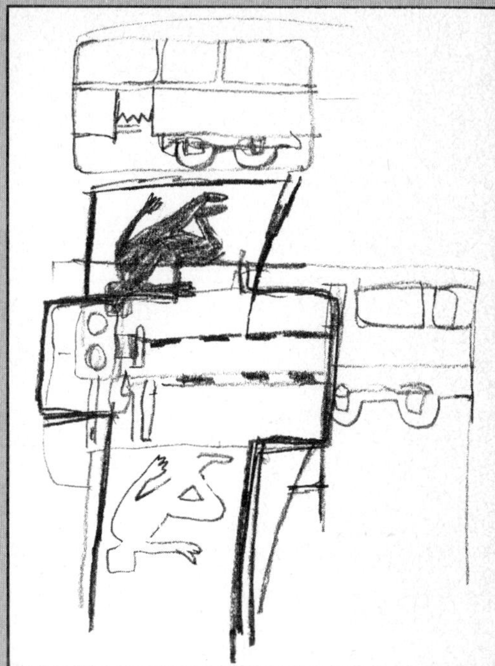

Requirement: Single 12 in. × 9 in. caption with lettering.

First pencil roughs: Road and train, railway lines, gates, smashed car – perspective makes it confusing – add crumpled figure – too confusing.

Next rough: Train head-on, crumpled car.

Next rough: Plan view of road and rails – looks like a cross – cross/death (symbolism) crumpled figure, lights and barrier – where shall I put the train?

Final rough: Large irregular cross with figure, lights and barrier – part of train growing out of cross – line-up part of carriage with barrier – striped barrier, like symbol for railways on maps. Modify the shape of the cross to make more room for the lettering – arrange as four lines – sans serif letters.

Caption: Cut cross shape from white paper and stick to black 12 in. × 9 in. card – cut figure from piece of photocopying paper and stick on cross – cut piece of light grey card for train – line up with edge of cross – draw lights and barrier and windows on train with felt-tip pen – paint wheels white – tone down white cross with yellow instant marker – makes it look like lichen – work out size of lettering – do spacing rough – then apply Letraset.

These are roughs for the title caption on page 80; the text explains clearly how the designer developed his idea from the brief to the finished caption.
To some designers the roughs above may appear to be very casual and not very explicit, and they do not use any of the materials in the final caption. The roughs are not even done on 4:3 ratio paper. This is just one designer's method; others would approach the problem in a completely different way and use different materials. In this case, the author nearly always works this way, as he is concerned in finding the right elements, thinking two steps ahead of what the pencil is doing, selecting the materials, scale and tonal range. Through experience, ideas are naturally formed in a 4:3 ratio without the frame being defined, so many things being instinctive that it is often difficult to explain why one does them.

The final caption which is the outcome of the roughs overleaf, is a bold design using four distinct tones and mainly cut-out shapes, plus a small amount of drawing. It relies more on symbolism than explicit detail, which the programme will provide.

Time taken: Fifteen minutes for the roughs – forty-five to make the final caption.

This is just one way of tackling the problem. Another designer might have used the same components, but the final arrangement would be different, probably without drawn imagery. The choice of the method of production (drawn or photographic imagery) is usually suggested by the producer, but it is often made in a general discussion between the designer and the producer. Obviously the designer can try to persuade the producer that a particular style and method of titling would be appropriate for the programme.

Whatever the individual method, all titling captions and sequences rely on the methods and styles of production outlined in the previous chapters.

HAND-DRAWN TITLE CAPTIONS

This form of title caption provides the designer with tremendous opportunity to develop some highly original visual imagery, covering a wide range of styles and subjects. Because such titling can be produced quickly and inexpensively, it is often chosen by the producer. It is also extensively

The promotion slide (*left*) has a bold, vigorous style coupled with good drawing and an economic use of tones. It is the sort of caption that needs painting quickly and directly to retain spontaneity and breadth of drawing.

This caption (*right*) has rather a flat form of presentation, which is directly linked to the mood of the programme. Notice the more subtle drawing and the varying scales of the elements.

used for programme trailers and commercial breaks. But because it is used so much, ideas and methods soon become familiar, and there is a continuous search for new ones. This in itself makes it a demanding and stimulating area in which to design, and pushes the designer's creative ability to its limits. Also, the extremely wide range of subjects and programmes presents a bold challenge and is of great interest to the designer, who is far less restricted than he would be if committed to only one type of programme. Variety makes the designer keep his ideas fresh and creative.

The simplest and quickest form of hand-drawn title caption, in terms of materials and techniques, is that produced by drawing with black felt-tipped pens and painting with designer's grey gouache. The only problem is that the captions need to be clear and direct; the idea must be well formed and clearly stated to retain its freshness. The best designs contain just a few elements and are probably produced with two or three well-defined tones. The caption for *The Wages of Fear* is a good example of this; it is expressed in black, dark grey, light grey and some white plus some linear drawing with a felt pen. It depicts the two main characters of the film, and the lorry loaded with explosives which is the chief prop of the story. This design has spontaneity and strength, combined with complete control of the materials used, including the positioning of the lettering in relation to the image. The style of drawing is strong and forceful, and is complementary to the rough, tough story in the film.

The smoother, more sophisticated style of *Portrait of a Loser* would not have suited the previous example: the subject of the programme is a smart, eligible young man, with all the right connections, who is always doomed to failure. The design shows the contrasting aspects of his character, plus a few events in his unsuccessful life. The use of flat grey tones helps to emphasize the flatness of this man's life. A fuller, more rounded style might well have conflicted with the content of the programme, and would have made the man more three-dimensional than he was portrayed.

A much more stylized presentation has been used for the *Caesar* title caption. It combines a strong brush drawing in

There is a strong contrast of materials in this design: brush drawing and rub-down mechanical tints, which reflect the conflict of the drama of vicious human relationships played out against a classical setting. Note, too, the shadowy figure in the background.

This caption has an ambiguous quality in its imagery – a woman and her relationship with the two-dimensional man. The programme deals with the exploitation of sex for financial gain; the idea behind the design was the contrast between the fullness of the woman, and the man who represents all those who pay and are paid.

This is for a programme on the habitual criminal and the image suggests identity parades, and figures fitting into predetermined moulds. The design uses simple cut-outs made from negative photocopying paper which has been textured by rubbing with Copydex and toned with coloured felt-tip pens.

The basis of this design was the photographic reference of an ancient Egyptian wall-carving, which was produced as a simple brush drawing, and integrated with a stylized woman's face. The hair was drawn in coloured inks and then over-painted with white paint.

This design relies on the outline of the shapes and the simple use of colour. It hints at some of the associations of chemical warfare – blindness, paralysis and nuclear radiation.

The initial element in the production of this caption was a blue-cast colour photograph which was cut and mounted on a black background. The figures were drawn on top with coloured inks and white paint, but some of the texture of the photograph shows through, thus helping to unify the elements. The lettering is hand-drawn. The photograph of grass was used because of a recurring theme in the story.

The drawings of the models were based on a series of photographs taken by the designer, which have been translated into bold shapes and fine line. There is an intended negative quality in the face on the left of the design, and collage imagery suggests the conflict between the model and the product.

The components of the design on the right are two colour-tinted black-and-white photographs which have been mounted on coloured paper. The slightly bizarre imagery suggests a drug-induced dream, with the staring eye as a mirror reflecting or seeing through us.

Driving and drinking is rather like encapsulating a car in a bottle of alcohol; who is driving the car, the driver or the alcohol? Is it alcohol on wheels? The idea behind this caption was to give the viewer a visual jolt without showing the horrific side. The association is surrealistic, but perhaps it will have more effect on the viewer than accident photographs.

black with rub-down mechanical tints. The variation in the tints represents the classical colonnades of the setting of the play, the drawn figures the brutality of the story, the murder of Caesar and the hollow oration of Brutus. A similar style has been used for *Sex Game*, except that the figure of the woman has been constructed from pieces of a photograph of a face, and then printed with a coarse half-tone screen. The figure is made up of five separate portions of the photograph which have been cut out and carefully matched. It has an ambiguous quality, because it could be a front or rear view of the woman. The man contrasts strongly with the woman; he is a two-dimensional character whose business is the exploitation of sex for financial gain. The woman represents every female to be found in pin-up magazines, clubs and strip-joints; the lettering reflects the voluptuousness of the subject.

Quite often photographic references are useful in the development of a design. In the caption *Homage to Cleo*, part of the design was based on a photograph of a detail of an Egyptian sculptured tomb. It is not a faithful copy of the photograph; certain forms only have been used in a stylized brush drawing and then integrated with the face of a woman. The decorative treatment of the head was achieved by using a cut-out piece of white paper which was drawn with coloured inks and finally over-painted with white paint to emphasize the swirling pattern of the hair. The choice of lettering helps to reflect the classical style of the design. The yellow background of this caption would reproduce as a light grey and the red a mid to dark grey, when shown on monochrome television.

The caption for *Three Blind Mice* makes use only of cut-outs. The three figures are cut from some negative photo-copying paper and coloured with instant markers. The white graining round the edges is produced by smearing Copydex on the figures and then rubbing it off with a finger. This gently removes the photo-emulsion and reveals the white paper base. The cut-out shape on the left has been toned down with an instant texture. The programme deals with the habitual criminal, and the arrangement represents identity parades and people being handcuffed together. The black silhouette on the left signifies that one of the figures, who will fit back in that place, is bound to be caught again for some offence.

In *Chemical Warfare*, cut-out shapes have been used again, but there is more drawing, although the design relies mainly on the outline of the shapes and a simple use of colour. The content of the caption hints at some of the possible effects of chemical warfare – blindness and paralysis – and the associated effects of nuclear radiation. The use of bold shapes with an absolute minimum of fine drawing can be a good combination for some captions, like *A Nice Girl Like Me*. The basis of this design was a colour photograph of grass (but printed with an over-all blue cast), which was cut

These two title captions for educational programmes are simple and decorative in their treatment, give a strong idea of the content of the programmes and make attractive openings.

to shape and mounted on black. The top portion has not been painted, whereas the girl and three faces have been drawn on top of the photo with coloured instant markers and some white paint. Some of the texture of the photo still shows through the colour and unites it with the top portion. The lettering is hand-drawn and painted. The subject of the programme is a young girl who finds she is pregnant but does not know who is the father.

The Anatomy of a Model uses a collage of bold shapes with some fine line drawing. There is an intended ambiguity in the faces enclosed in the body shapes; models are faces, their function is to advertise or promote clothes, and a conflict is created between the face and the product. The orange background would reproduce as a mid grey on monochrome television.

Two examples of title captions for education programmes are *What is a Tornado?* and *Intensive Farming*. These are simple and straightforward designs drawn in black on mid-grey card. They are both decorative in treatment, but give a strong suggestion of the content of the programme and make an attractive and informative opening.

Many of these title captions could also be used for promotion stills and commercial breaks. They involve a range of styles and materials, and some of them can be produced very quickly once the design has been worked out. The time taken to work out the design depends upon how much reference is needed and how much proves unfruitful. The important thing to remember is, first, to get rid of preconceived ideas and, second, to make sure that the imagery is compatible with the content of the programme.

PHOTOGRAPHIC TITLE CAPTIONS

In some respects photographic title captions are much harder than hand-drawn ones for the designer to control. This depends upon whether the designer takes and prints his own photographs or instructs a photographer to do it for

him. When the designer does it all himself it is possible to modify the idea, either in the shooting stage or, more so, in the printing stage. Certain ideas would not be thought of when somebody else is doing the work, and this rules out the possibility of good effects being achieved by accident. This is something the designer must not rule out, for even if the result cannot be used in the job in hand it might be suitable in another context.

In general the photo title caption will take longer to produce than the hand-drawn caption, because it is unlikely that the required negative or print is immediately available. It will either have to be shot specially or a print will have to be made from a negative borrowed from a photo library. The designer must have a clear conception of what he wants at a fairly early stage in the design process; there is no time for second thoughts.

Some designs will contain only photographic elements, but they need not be produced exclusively by photography. The caption for *The Building Crisis* uses a simple photo image of a brick wall, which has been cut and torn, and then arranged into a design which evokes the feeling of crisis and break-up.

When photographs are used in collage (cut up, torn and overlaid), they are best printed on matt single-weight paper, which is easy to handle and cut and which can be easily retouched, painted and drawn into. Photo-copied documents can also be used, but the emulsion is not so receptive to retouching.

Drop Out Dream uses a combination of materials. The main element is black-and-white photographs, one of the girl's body and frame and the other of part of a man's face. These have been tinted with coloured instant markers and mounted on some coloured and toned paper. The idea behind this caption was to produce a startling but realistic image of a drug-induced dream – the single staring eye that looks through us; we cannot hide from it.

Like drawings, photographs can be reduced to a simplified statement, to the basic elements of the subject. This is suitable for subjects such as *Death Just a Job*, a programme dealing

The photographic elements of *The Building Crisis* (*left*) could not be simpler. The picture's impact relies entirely on the arrangement of the bits of brick wall and lettering, and the association of ideas.

The photo design on the right has a stark quality which is achieved by making copy negatives on line film, thus reducing the normal tone photograph to harsh black and white. This caption merely establishes the identity of the man; the programme will reveal everything else.
There is always a danger of the designer's being too explicit when producing imagery for titles, and taking away something of the impact of the programme. There is a thin dividing line, although the designer is guided by the producer's brief.

The photographs of cars (*above*), apart from being printed in varying scales, have been simplified by painting some areas mid grey. The hand-drawn lettering suggests signs painted on roads.

The imagery of *Homicide* (*above right*) and *Murder File* (*below*) is a collage of photographic elements and drawing. In the first, the images are fairly realistic and straightforward, whereas the photographic elements in the second are used in a more abstract way. Notice the varied scale of the images, and the tonal relationships.

with an S S officer who was involved in the death of prisoners of war, and who became a normal citizen with a state-issued identity card after the war. It could be argued that images of his past should be shown, but such imagery would reveal too much about it and might be too strong for the opening of a programme. It is better to use imagery which is stark and slightly sinister; the programme will reveal the rest.

The car images of *Motorway* have been made very contrasty and then slightly modified by painting some areas with a mid grey. Varying scale and perspective of the images, and hand-drawn lettering reminiscent of road signs, are used effectively. Whereas the cars in the last example are still separate images, the photographic images in *Homicide* are much more integrated. This caption started as four individual photographs. The print of the prison officer was copied by painting on tracing paper and then dry-mounted on the still of the man's face, so that the face could be partially seen through the tracing paper. These were mounted on a black card along with the prints of a revolver and a police 'mug shot'. The prints were then modified with black and white paint, so that there was more continuity of shape and details. The face partly obscured by the tracing paper was revealed by painting the light areas white, thus creating a misty image of the man's eye through the officer's face. Again the scale of the separate images is very important; this idea could not work if all the images were the same size. The stark black-and-white nature of the images was important to the conception of the design.

The next caption, *Murder File*, is of the same genre, but the photographic elements are less complicated. Pieces of photo-copied images have been cut up to represent the legs and skirt of a victim, and tinted. The importance in this case of the photograph is not to use the images in a representational sense but to make them something else; the tree branches become part of a body, and trees and blue sky represent freedom, but the body is dead – the contrasting imagery serves to emphasize this fact.

It is not unusual to see a ship in a bottle, but rare to see a car in a bottle. Yet driving and drinking is rather like encapsulating your car in a container of alcohol. Who is driving the car, you or the alcohol? Is it alcohol on wheels? This was the idea behind the *Driving and Drink* caption, an image that would give you a slight jolt without showing the horrific sides of the subject. The bottle was a black-and-white photograph coloured with instant markers; it was necessary to obliterate recognizable trade marks. The car was a cut-out colour photograph. The bottle could have been a colour photo, but it was decided to use a hand-coloured print in order to give the image a surrealistic quality. The same surrealist quality is to be found in the *Defection* caption, with a face appearing out of a man's ear and a hammer and sickle on his eyeball. The programme was about spying and the defection of spies.

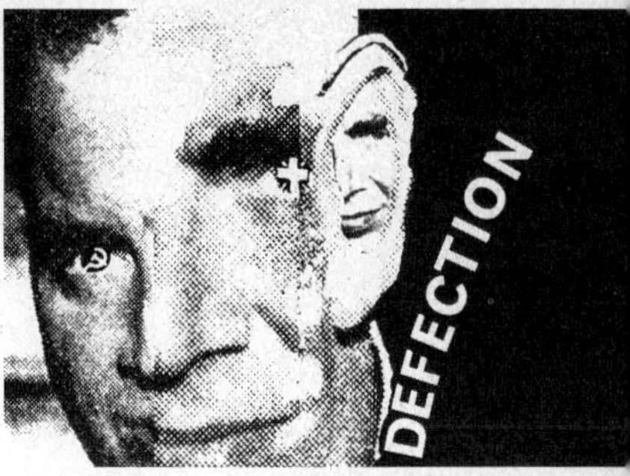

Again a photo collage which makes use of half-tone screens and exaggerated scale.

Most of the examples mentioned have made use of photographic elements but have included some drawing, painting and rearranging to make the finished design. Colouring by hand instead of using colour photographs was done to achieve certain textural and surrealist qualities. However, there is a means of electronically injecting colour into a black-and-white print with a colour synthesizer. If a subject is photographed against a black background, and is of a contrasty nature, then it is possible to change the background to a particular colour, and to generate a colour for the subject by using inlay as well as a synthesizer. The colour has a flat continuous tone, which is superimposed on the image, rather like placing a piece of coloured tri-acetate on top of a photograph. For some title captions this method can be very successful; only black-and-white images are initially necessary, thus saving time and expense.

Most photographs used for title captions are in a sense very stylized. They are specially shot for the design, or are just a portion of a photograph from a library, or possibly are cut out from the original background. Normal photographs are not usually suitable because they have too much detail or the background is too fussy. Nearly all the photographs used are shot against a black, white or coloured background and the lighting is arranged so that there are no irritating shadows or reflections, unless they are integral to the design. The reason for this is to eliminate all unwanted information and to concentrate the viewer's attention on the subject of the caption. A title caption or promotion still is only on the screen for a few seconds, and nothing must be allowed to interfere in its communication with the viewer.

TITLE SEQUENCES

Title sequences are only an extension of title captions, the major difference being that the design information can be built up over fifteen to sixty seconds rather than appearing all at once on the screen. The sequence introduces the elements of time and movement, and integrates sound. Movement can be defined as the juxtaposition of images with time,

The simplest form of title sequence consists of two to six 12 in. × 9 in. captions, each with a related image and some lettering. These two captions use stills from the programme with a simple arrangement and clear lettering.

and involves the length of time the image is on the screen, and its position, scale and tone. The movement can be real or implied, that is, movement of an element of the image or movement created by cutting and mixing from one image to another. Real movement is likely to be live action, whereas implied movement is stop-frame animation (see p. 96).

The range of imagery is limitless, but because a sequence demands a progression of images and ideas much more work is required from the designer. He must work out an idea as a sequence related to the subject and perhaps to some music, and then produce the necessary artwork or photographs. This will take considerably more time than merely producing a title caption, and will involve much more research.

The simplest form of sequence would consist of two to six 12 in. × 9 in. captions, each with a separate but closely related image or images and some lettering. The director would cut, mix, wipe or defocus between each caption, probably timing them to the phrasing or beat of some music or sound effects. There would be a synchronized build-up of information in the sequence, perhaps depicting some of the characters in the programme or evoking a mood complementary to the subject.

The two captions for the educational programme *English Literature* use images from the content of the programme, stills from a dramatized excerpt, and some imagery connected with the subject of the poems. The link between the two captions is the woman's face; the arrangement of the images is very straightforward and clear and the lettering is given its correct importance.

The four captions for *A Plastic World* were the result of a photo session in a studio, when about thirty exposures were taken in order to get the right poses. They were intentionally taken against a black background, and the black areas were burnt-in when printing. The model was dressed in a plastic mirror and all the lighting was by electronic flash. By mixing between each caption a simple form of animation was achieved.

In order to heighten the impact of the sequence for *The Wall*, inter-negatives were made on Kodalith translucent film, which makes the photographs very contrasty and coarse, helping to emphasize the harshness of the subject. The sequence starts with a close-up on the centre of the

The four captions for *A Plastic World* (*opposite*) were the result of a photo session with a model in a studio. About thirty exposures were shot (the contact strip shows some of them) in order to obtain the final four captions. When working with a model it is not unusual to shoot quite a lot of photographs, because if she is moving, not every shot will be perfect. There is bound to be some detail such as facial expression which creates the wrong mood. Be prepared to shoot plenty of exposures so that you can choose the best photographs for the sequence, without having to make do with shots that are not exactly right for continuity in the finished result. By mixing between each caption a simple form of animation was achieved, and provided an effective opening for the programme.

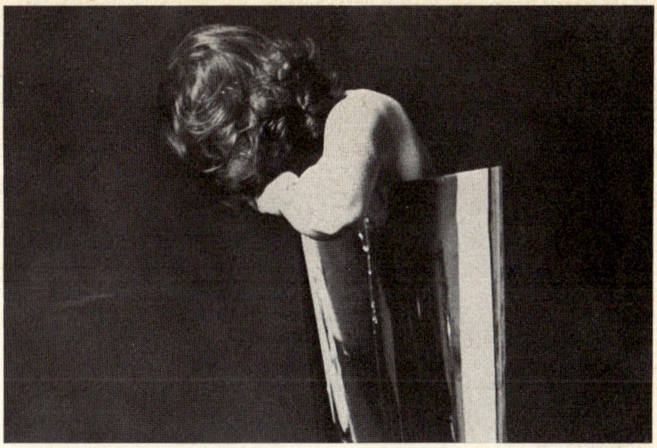
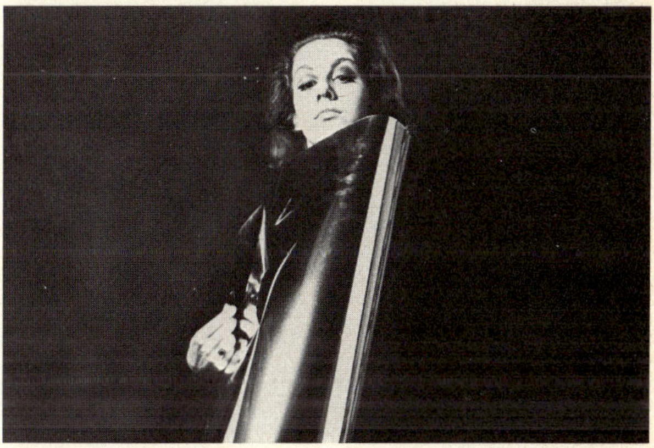

THE
WALL

first still and a quick zoom-out to the whole caption. This cuts to a pan left to right on the second still, cut to pan down the church tower to the wall in the third, cut to the whole of the final still and zoom-in to the lettering. The stills were televised on caption stands in the studio; but they could have been filmed in advance.

The series of captions for *The Mark of Glory* makes use of a free style of drawing with a soft pencil, which is very expressive. The reason for using this instead of photographs was that photographs had been used many times before for the titling of plays and documentaries dealing with the Second World War, and therefore the images were very familiar. Using them would have immediately 'labelled' the programme and perhaps started it off on the wrong foot for many viewers. Also the photographs would have presented too much of a documentary image when the programme was a play, though based on some historical facts. Although the drawings have a certain strength, they also evoke a feeling of fragility, of human frailty and emotion, which was an important element of the play. The original drawings were quite small; they were copied on line film and enlarged to 15 in. × 12 in. to emphasize the quality of the line and to accommodate zoom-in on some of the captions, so that the lettering could be superimposed. This lettering was white on black on separate captions, and was phase-reversed so that it could be inlayed on the almost white background of the drawings.

The four photographs above, printed from Kodalith inter-negatives, provided the twenty-seconds sequence indicated by the stills opposite. This was achieved by zooms and pans over the photographs, which were printed 15 in. × 12 in. to accommodate the movements of the TV camera. The movements were: first still – close-up of centre, zoom out, cut to second still, pan left to right, cut to third still, pan down, cut to fourth still, slow zoom-in and hold.
The harsh, grainy quality of the photographs gives impact and is appropriate for the subject.

1
THE MARK
OF GLORY

2
THE MARK
OF GLORY

3

4

5
BY
PAUL
SCOTT

6

7
WITH
EWAN
HOOPER

8

9

10

11

12

This sequence was produced from eight pencil drawings, which were photographed on line film and enlarged to 15 in. × 12 in. This style was used, instead of war photographs, because it was much more expressive and in sympathy with the content of the drama. The drawings have a harsh strength, but they also evoke a feeling of human frailty and emotion which was an element of the play. On drawings 2, 5, 7 and 10 the camera zoomed in to accommodate the title lettering and credits. The black lettering was produced from white letters on black on separate captions, which were phase-reversed and inlayed on the drawings, so that they would register on the light-toned areas.
Once the research was done, the drawings were completed in one day, and it took only a few hours to copy and print them.

Electronic Circuits and *Human Physiology* are further examples of stills and styles that can be used for a series of captions for titling. The transition between the images in the first example was by a wipe from top to bottom of the screen; the second used a slow mix from one image to the next. Most of the examples quoted are fairly simple designs, where the information or mood is progressively built up by straightforward means. No complicated techniques or facilities were needed, and the designs could be produced quickly once the idea was formed. All the drawings for *The Mark of Glory* were completed in one day once the necessary research was done, and copying and printing them took only a few hours.

FILM ANIMATION

The next type of titling sequence, film animation, is not cheap to produce, and it takes considerably longer to work out the idea and produce the artwork. This is also called stop-frame animation and it is filmed on 16-mm. or 35-mm. cine-film.

There are several methods of producing artwork for animation, some relatively simple, others exceedingly complicated and very time-consuming. It is sometimes necessary to have an army of assistants to cope with all the drawing, tracing and painting. Apart from producing the artwork, the designer must also give very explicit instructions to the rostrum cameraman. In some small outfits the designer has to be his own cameraman, so that he must be skilled in operating a rostrum camera.

We shall first consider the design operations which are required before filming starts. Once the idea has been mooted that an animated sequence is to be used for the titling presentation, there will be at least one meeting between the designer and the producer to determine the form of imagery, the length of the sequence and its aim. The sequence might be animated letter forms, a self-contained unit, or a direct lead-in to the programme introducing some of the personalities or the opening of the story. It can also be used to evoke a mood or setting for the content of the programme. Many of these decisions are arrived at by the producer and designer throwing ideas around until there is a basis for a brief, so that the designer can do some research and work out an idea for a storyboard. The storyboard is the main way in which the designer communicates his ideas for the sequence to the producer, before starting on the artwork. It is an image–time graph, and shows a series of static images which indicate the content, the form of imagery, the type of movement and the style of the sequence. The storyboard is like an architect's plan; it shows the basic structure and maybe a few details, but it does not show how many bricks are

The designs for *Electronic Circuits* (*left*) and *Human Physiology* (*above*) are roughs. They use simple imagery, but it is the cumulative effect of the sequence, not the information on each individual caption, which is important.

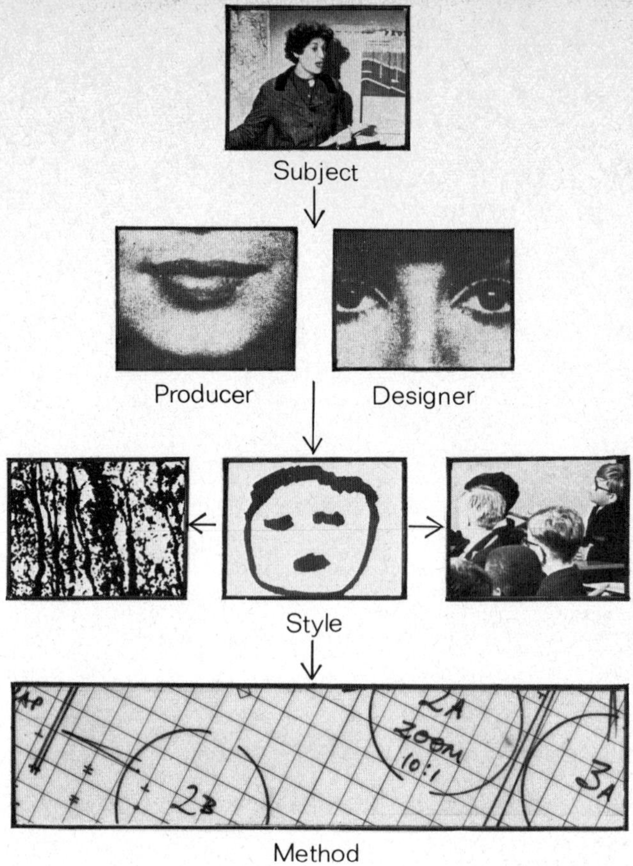

This diagram indicates the stages in the development of a titling sequence, starting with the subject of the programme and followed by the discussions between the producer and designer to determine the brief. The designer then does his research, forms some ideas and works out the style, and then decides which method of presentation (film or studio) is the more appropriate for the finished result.

going to be used nor their texture. The style of the storyboard varies from designer to designer, and also according to the type of imagery to be used in the final product. Some might use simple line drawings or bold felt-tip pen or brush drawings, others might use pieces of photographs or photo-copies pasted up and painted. The storyboard may not show the exact images required, but it gives an indication of the idea. Whatever method the designer uses, it is important for the storyboard to be clear and communicate the idea quickly and without too much explanation. There is no point in spending a long time making a storyboard if the producer does not like the idea. Its design must come quickly and naturally to the designer; economy is the key factor.

In some cases, the theme music or sound effects will have been chosen before the storyboard is started, and there will be some indication on the storyboard as to how the sound synchronizes with the images. Generally the sound is not chosen until after the storyboard stage, but in either case it is the key to the next stage.

With most animated sequences it is the music or sound effects which determine the rhythm of cutting, the length of shots or movements, so that the phrasing of the music synchronizes with a change of shot or the introduction of a new element. The sound is recorded from discs, tape or live material on to 16-mm. fully coated magnetic film; then,

using a synchronizer with sound head or an editing machine, a careful analysis of the phrasing, bars, beats and off-beats is made by marking on the magnetic track with a chinagraph pencil. These marks are related to the frames of the track and a chart can be drawn up which shows, in frame numbers, the changes of shots and all the necessary cues to which the film has to be synchronized. This chart is important in the preparation of the artwork and of the 'dope sheets', which are the filming instructions given to the rostrum cameraman.

The storyboard above is a quick and informative way of explaining the contents of a sequence to the producer. It indicates the form of the imagery, but not necessarily exact detail. Note the instructions and timing for each image.

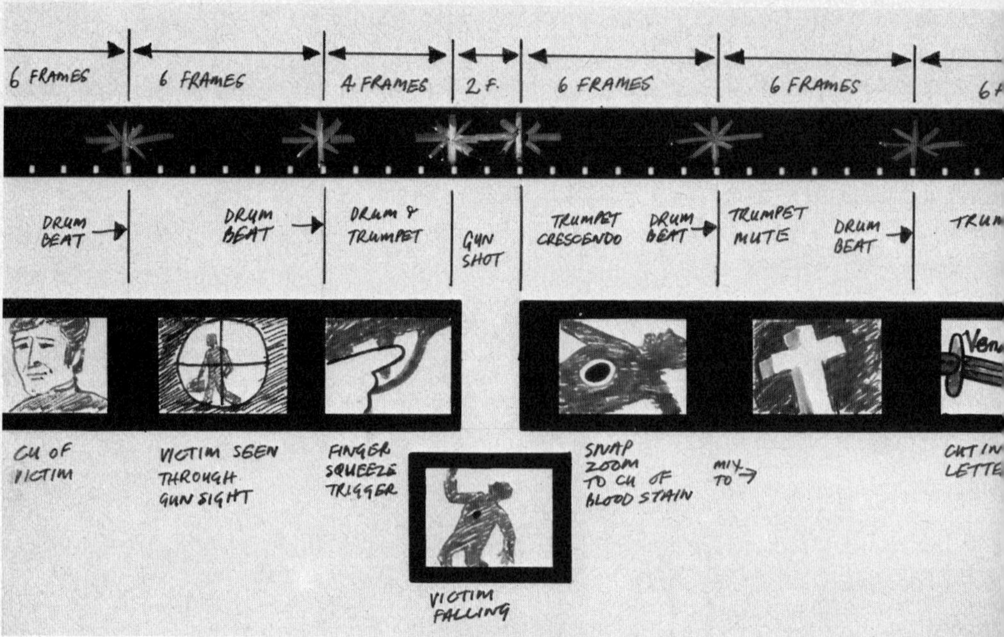

The diagram above shows how the sound for a sequence is analysed and related to the artwork. Very exact timings are made so that the designer knows how many seconds or frames each image and movement requires, so that the film will synchronize with the sound-track.

The sound chart or 'breakdown' will tell the designer how many frames a particular movement can last, and whether he should use one drawing every frame or two frames and how much the image will move on each drawing. If there is one drawing per frame, then twenty-five drawings are required for every second of film (TV film projection speed is 25 frames per second European standard and 24 frames per second for American standard).

When still photographs are being animated, the designer should determine how far the camera can pan or zoom in a given number of frames or how many stills are required for a phrase. Some of the working out of the drawings and movements involves simple mathematics; frame numbers are related to how far the camera has to move in inches or centimetres, and this information is also used on the dope sheets.

All artwork and photographs for animation are produced to standard sizes called 'field areas'. This is the area which the camera records when set at a specific position. Field areas normally range from field 2 to field 15, the numbers corresponding to the horizontal measurement of the area in inches; thus field 3 would be 3 in. $\times$ $2\frac{1}{4}$ in. These fields are directly related to how near or far the camera is from the artwork – the larger the field the greater the distance between camera and artwork.

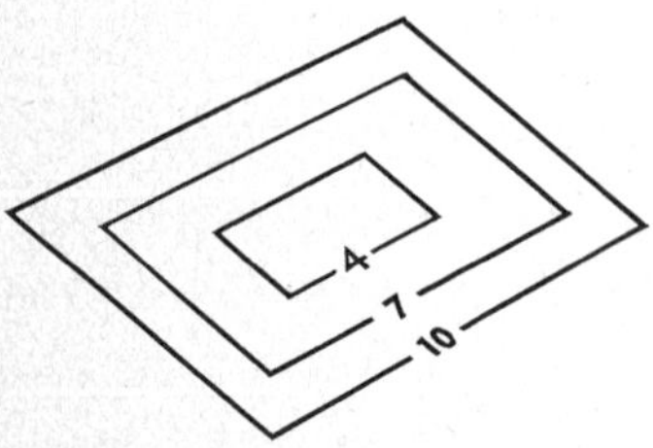

All animation rostrum work is controlled by 'field areas', which are an indication of how far the camera needs to be from the artwork. On the chart above, the higher the number the larger the artwork.

A rostrum camera always operates pointing down on to the artwork, which is placed on a horizontal table (panning table), therefore the camera is operating in a vertical plane. The panning table is geared to move in a north–south and east–west direction, controlled by hand cranks or electric motors. The camera is positioned in a cradle which is at

This is a storyboard for a live action sequence. It uses a very lively and free style of drawing which, coupled with the written instructions, gives a very clear indication of the action and associated sound-track. In most cases the designer produces only the idea, and the live action filming is done by a film cameraman. However, if the designer has some experience of film-making – shooting, editing etc. – he will be able to produce a much more practical and meaningful design for the sequence.

This is a still from an animated film, based on an adaptation of the 'Nun's Priest's Tale' from Chaucer's *Canterbury Tales*. The film was fifteen minutes in length and was designed for children's programmes. The method of animation was rather unusual, because all the figures and animals were two-dimensional cardboard puppets with jointed limbs. The woman's arms, legs and head were joined to the body with flat split-pins, which are hidden behind. The animation was achieved by moving the limbs and other components a fraction of an inch every frame of film. In some cases the figures moved around the frame, as well as the limbs moving. All the other components such as the trees were also cut-outs which were stuck to coloured backgrounds. This form of animation gives slightly stylized movements (very appropriate for this subject) but it saves making hundreds of separate drawings on cel, to accommodate all the movements. It takes longer to film and requires considerable patience, but the results are worth while.

right angles to a vertical column and is geared to move up and down the column, thus giving different field areas. The animation camera has a fixed focal length, but a variable-focus lens, with automatic follow focus, so as it moves up and down the column (zooms) the artwork is always in focus. As well as the panning table being geared to move there is normally a set of movable peg-bars (to which the artwork is attached), which can be moved independently of the table by a geared hand crank. Most rostrums have a back-lit ground-glass panel on the panning table allowing cel artwork to be illuminated from behind, thus giving better results. Animation rostrums can range from the home-built 'lash-up' with a 16-mm. Bolex camera costing altogether £300–£500, to a custom-built rostrum and camera costing a minimum of £5,000, and, if extra facilities are included, possibly twice as much. The more expensive the rostrum, the more accurate it is and the better the results. Anything can be repeated, there are calibrations and counters to indicate all movements, and the process of filming is not so tedious because exposure is electronically controlled.

To help the designer produce the artwork in proportion to the field areas, a sheet of triacetate marked with the field sizes (graticule) is standard equipment. This can be laid over the artwork to check if all the information is within a given area, and can be used to plot pans and zooms. To make the lining-up of the artwork and the graticule more accurate, everything is punched with a special animation punch which cuts three shaped holes. These enable the designer to place all the artwork on a peg-bar, which corresponds to the punched holes, and then a series of drawings can be very accurately matched to each other as well as to the field areas. As already mentioned, the panning table of the rostrum also has these peg-bars, so that the artwork is correctly positioned for filming. Peg punches are expensive items (over £100), but they are absolutely essential for accurate registration of the artwork.

Animation implies movement – movement which is created from a series of static images and filmed one frame at a time. It can be achieved in two ways: first, by a single image on cel or card which is progressively moved across the camera's field of view by moving the calibrated table or peg-bar. The second method is to have a series of drawings on cel, registered on a fixed peg-bar; the position of the image changes on each cel, so that the image moves across the screen. The main difference between these two methods is that the first requires only one drawing, whereas the second needs a minimum of twelve drawings for each second's worth of film. The advantage of the second method is that information can be added as the movement progresses, and it is also possible to repeat the sequence *ad infinitum* to achieve a continuous cycle of movement. Full cel animation requires a lot of artwork; it is often necessary to use two or three levels of cel and a background to accommodate all the

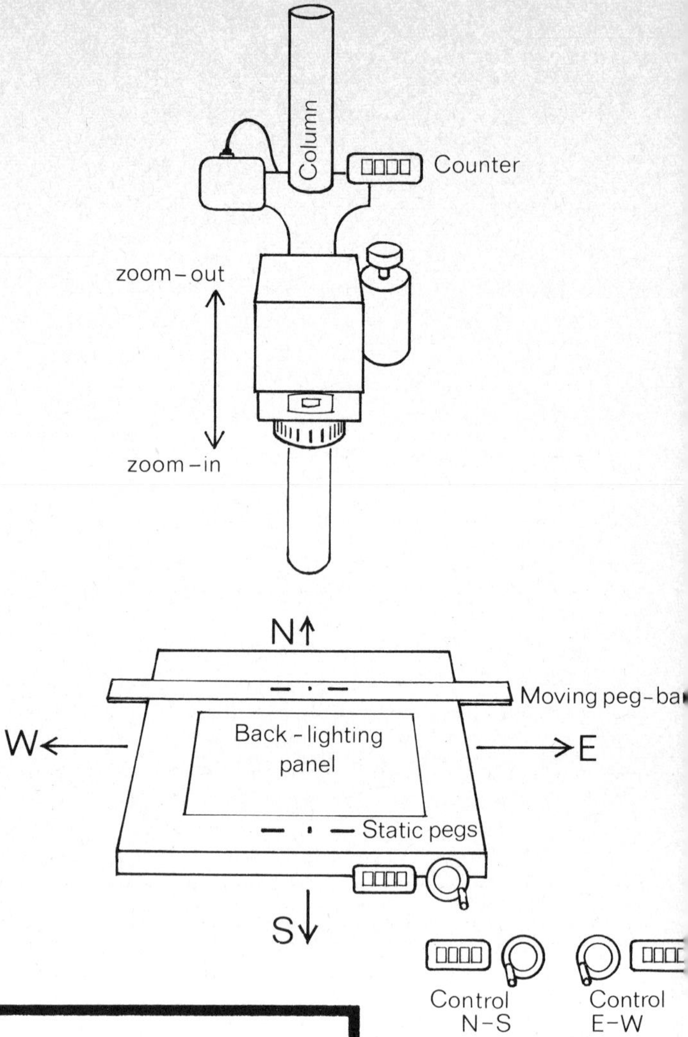

This diagram (*right*) shows some of the basic movements of an animation rostrum and camera. The camera moves up and down the column, thus changing the size of the field areas, and the panning table is geared to move in several directions.

The chart below is called a graticule. It represents field areas printed on a piece of cel, which can be laid on top of artwork and photographs to determine the field area required for filming, and put under the cels when they are being drawn, thus ensuring correct framing of the artwork. The holes punched at the top are for registration on pegs, which are used in the drawing and filming stage.

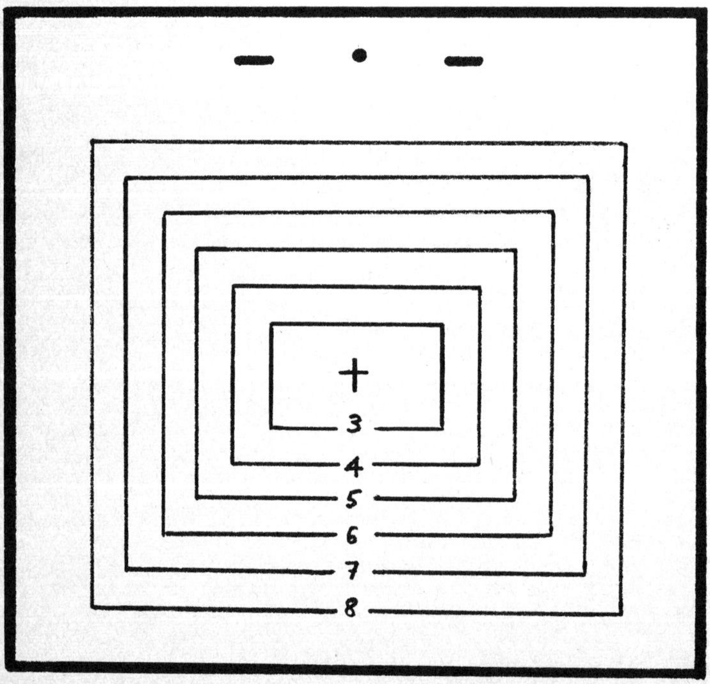

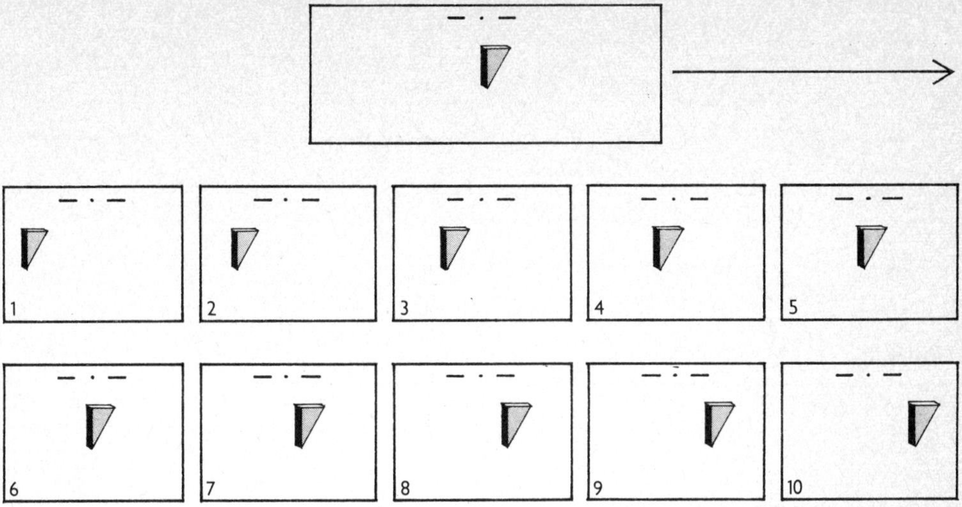

The diagram above shows the two basic forms of animation movement. The top line is one, where a single image is drawn on a long cel, and the panning table or moving peg bar is moved from left to right when filming. The other method is to make a series of drawings on cel, the object being in a slightly different position on each cel. When these are filmed, the panning table is static, the cels are changed every one or two frames (going from 1 to 10) and the filmed result is almost identical to the first method. The second method is used if the object changes shape while moving across the frame, and when other images are added during the movement.

movements. By using this method, however, complex and sophisticated images and movements can be used in a sequence.

Before outlining animation methods it is necessary to consider the motivation and planning of a sequence. Most sequences, because they are only thirty to sixty seconds long, have to present a story, feeling, mood or information in a very short time. This telescoping of time is unnatural, particularly in relation to the programme that follows, so the designer must make the sequence of images as natural as possible. This 'natural' quality is the secret of a good designer. He has control over more than just the elements involved, the style of drawing or photographs; he is presenting the viewer with a series of images and sounds which, by association of ideas, by logic or even ambiguity, evoke emotions and prompt the eye, ears and brain to certain definite conclusions. These conclusions or responses operate at several levels, some logical and others illogical, some aesthetic and others material. Also there is the factor of interference. The viewer is not captive: someone can talk, the doorbell might ring or the baby cry and disturb his concentration. Of course the designer cannot cope with all these eventualities, but he can try to shock or seduce the viewer into watching, by the use of effective images and sound. He must convince the viewer that the programme to come is worth watching, whether it is a play, a comedy show or an educational programme. The designer is not only creating the ambient mood for the viewer, but he is also presenting the image for the content of the programme and its performers.

The planning of an animation sequence is more straightforward. The main problem is finding the right combinations of images and their associations, the style to be used and the appropriate timing. Ideas cannot normally be pulled out of mid-air; the clue to the starting-point is research. The script prompts, the designer searches, the reference suggests, the designer manipulates, draws and designs, and reaches certain conclusions. This process can take a long time and many sheets of paper, 'paper' being the operative word. Just thinking is not sufficient; ideas and points of reference must be drawn, arranged and moved around on paper, perhaps only to be rejected and a fresh start made. That is the process; exactly how each designer thinks and doodles is variable, but there should always be an analytical stage of assessing the strength and validity of the ideas and reaching conclusions, so that the ideas can be brought together to make a unified statement and a storyboard. When storyboarding the idea, decisions need to be made about the style of the images, whether they are to be representational or stylized, hand-drawn or photographic. Stylization includes characterization and period flavour. Are the images meant to represent today, a hundred years ago, or the indeterminate future? These considerations are important in making the sequence convincing and natural and an integral part of the programme.

Once the idea has been worked out and storyboarded, then the designer can determine what form of animation is required. The simplest method of animation is the use of cut-out shapes which can contain drawn or photographic imagery or make use only of colour and tone. These are manipulated by hand on the panning table, each shape being moved a fraction of an inch each frame. This method is ideal for simple movements using collage-type imagery; additional movement can be added by tracking the panning table and zooming in and out with the camera. The limiting factor with this method is that it is often difficult to move the shapes smoothly and work out exact calibrations. The animation tends to be rather crude, although it is very good for diagrammatic sequences. A refinement of this method is to stick the shapes on cel, so that they can be moved by the calibrated peg-bar. It could even change shape if there is a series of cels, each with a carefully matched new cut-out shape. Although this is a simple method of animation, the results can be surprisingly good and effective. This type of work is ideal for diagrams and programme inserts and the artwork can be produced very quickly, which is a great advantage. So long as the shapes and artwork (drawing, etc.) are well designed, the visual impact of the images and movement justifies using this method.

The full cel method requires totally hand-drawn imagery, although it is possible to achieve a similar effect with a series of very carefully matched photographs. In producing the artwork, it is necessary to make a careful analysis of the movements and find the minimum number of separate cels required. For example, if the subject of the animation is a man walking from left to right across the screen, the body and head of the man would be on one cel, while a series of cels would be necessary to cope with the walking action of the legs and arms, and some movement of the eyes. A cycle of 25 cels is needed for a second's worth of action and they would be repeated 1 to 25 starting again at 1, continuously, while the peg-bar is moved across the panning table along with the body and head of the man. If any variation is needed in the action of the man, or if new elements and images are to be added, additional cels are necessary. An alternative to this method is for the walking man to stay in the centre of the screen while a moving background, going from right to left, creates the illusion of the man walking past the background. The body and arms and legs cels would be on fixed pegs and the background on moving pegs. The panning table would not move. Again, any new images to be added to this scene require new cels.

When drawing on cels, the outline is drawn on the front, but all solid areas and colours are painted on the back, so that the colours are more full. Special cel paint can be used, but a good alternative is plastic emulsion paint (slightly watered down), which is cheaper and easy to apply. Gouache paints can be used, but they need mixing with a little soap so that they adhere to the cel properly. Sometimes the images need to be semi-transparent and luminous; this effect can be produced with coloured inks or rub-down instant colour. When applying paint on cel, use firm strokes with the brush, otherwise it will not give a flat result.

Photographic images can be combined with drawn imagery with effective results. The photos need to be printed on single-weight paper so that they can be cut out easily and stuck on the cel. A considerable number of programme titling sequences are produced with combined photographic and drawn images. Sometimes the photographs are printed on film instead of paper, so that the images are semi-transparent.

When producing the artwork for animation, the normal method is to draw the key positions (keys show the main difference in shape or position of the image; a key drawing is roughly every fifth or sixth drawing in a sequence) and some of the in-between positions on punched drafting paper. All the details can be worked out in these drawings before the cel is placed on top; the drawings are traced off, one at a time. If the sequence is very long, the designer will do the key drawings and perhaps some of the in-betweens and work out a chart showing all the colours and tones to be used. These would be passed on to the design assistants who trace and paint the cels. This leaves the designer free to work on other aspects of the production. If the finished cels are to be backlit when filmed, then each cel must be checked over a light desk to make sure there are not any pinpricks of light showing through the paint. Backgrounds can be painted on cel or on thin card, depending on whether backlighting is used, or they can be made up of tissue paper and coloured cel. As far as possible backgrounds need to be simple, so that they do not swamp the scale and importance of the main images. For many titling sequences, the backgrounds are

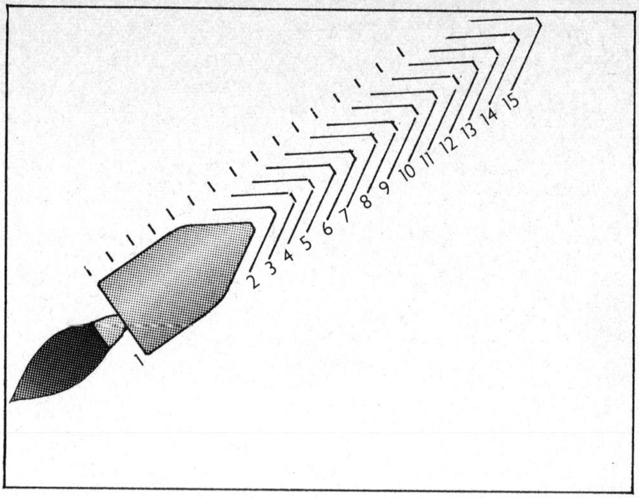

When planning the artwork for animation, the normal method is to produce charts such as this, which accurately shows the position and movement of the rocket for fifteen separate cel drawings. It is not always necessary to chart all the movements, if the movement is regular. Key positions only are drawn, such as 1, 5, 10 and 15, and the in-between positions can be drawn directly on the cels. Apart from charting the movements, these charts are placed under the cel, and the drawing is traced and painted on the cel.

plain colours or black, but sometimes textured backgrounds are appropriate. The priority for the designer is to be sure that visual information and lettering can be understood and are not fussy.

There are two kinds of still-photograph animation. In the first, the images exist only in the form of still photos, engravings or paintings, either because they date from before the era of cine film or because the scene was not thought worth filming at the time. The second kind consists of sequences specially designed as animated stills, the photographs being taken for this production only. The second is often used because it is cheaper than covering the same sequence by live action film, but it also has many advantages over live action film. The film can be edited in the camera as it is shot, mixes and fades can be incorporated at the same time, and because of the stylized form, action can be squeezed into a shorter time without any loss of continuity. For some programmes an animated still sequence can be the best way, artistically, of creating a different mood (for a flashback or a dream, for instance), and it can integrate well with live action programme material. Also it is possible to change the rhythm and pace of the sequence very easily, and to film very accurately to a predetermined sound-track.

When filming old photographs and prints it is possible to view the whole print and also to examine very small details, changing shots by cuts and mixes, although too many zooms and fast pans in a sequence can be disturbing and detract from the value of the visuals. A black mask is useful for isolating or emphasizing a detail; it concentrates the viewer's attention and it can be used for placing lettering to obtain greater clarity. Animation of old prints is a very good way of creating the atmosphere of a particular period or event, either for a titling sequence or for a programme insert, and it is also a fluid way of presenting visual information.

These four stills from an animated sequence are a combination of hand-drawn and photographic imagery. Most of the movement was in the hand-drawn imagery, features such as the trees in the first still, which lose their branches, then change into hands and arms (third still) and two crosses in the fourth still.

The photographic imagery was printed on sheet film and some of it was semi-transparent, so that the drawing shows through. This imagery is mixed into the shot, and it pans and changes position with the drawings.

The drawings were brush-painted with black and white paint, and textured by scratching. They are part of a thirty-seconds sequence.

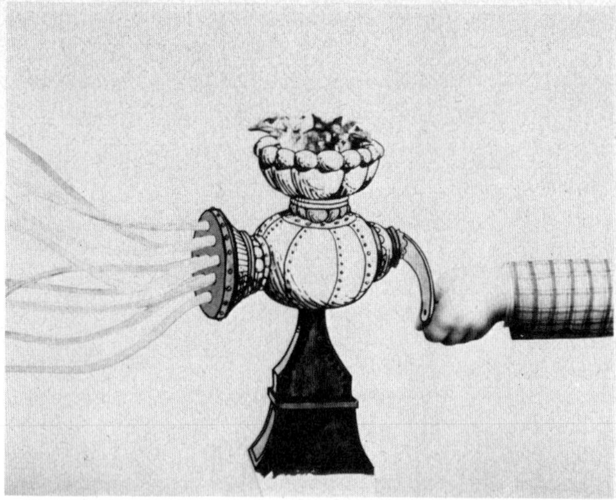

These three stills are from an animation sequence seen on *Monty Python's Flying Circus* which makes extensive use of photo-copies of old engravings and the painting *The Birth of Venus* by Botticelli. A large series of photographs are produced with the handle of the mincer in different positions, so that a cycle of photographs, when filmed frame by frame, presents a continuous winding movement of the handle and arm. During this action the camera pans across to the head of the Venus.

Sometimes photographs are taken with a still camera fitted with a motor drive, which in one burst can give a series of photographs which are almost like successive frames from a cine-film – one shot taken roughly every half-second. If these are printed carefully and matched, it is possible to animate them like full cel animation, two frames for every print. When shooting these photos, it is best to use a plain background, black, white or coloured, avoiding shadows.

When filming photographs, instead of punching the prints cut a number of strips of card, punch them, and then Sellotape them to the print. This stops the print from being mutilated and it could be used again for another job. Also make sure that each print or piece of artwork is numbered in the sequence to be shot. Once all the artwork and prints are completed, a 'dope sheet' must be written, giving instructions to the rostrum cameraman. The sheet shows the order of the artwork, how many cel levels there are, how many frames are to be shot for each piece, and how many frames there are for each shot, mix, fade and superimposition. It also tells the cameraman what field area to use and when to pan and zoom. These details must be very accurate, particularly when the film must synchronize frame for frame with a sound-track. Even if the designer is doing the filming himself it is still vital to make a dope sheet; memory is not good enough.

Animation rostrum filming is an art in itself. There are a number of books available which go into more detail, so we will only consider a few vital points. Most rostrums have a glass platen to hold the artwork flat; be sure that this is clean and free from reflections when lighting the work. If the counters on the panning table and peg-bars allow, reset them to zero before starting filming. Be sure to load the right film in the camera for the job in hand. Use a slow film (4 ASA) for completely black-and-white subjects, and a medium-speed stock (25 ASA) for normal tone black and white. Both those will be negative film but with colour there is a choice of negative or reversal film. There are two types of reversal film, one medium speed (25 ASA), the other fast (125 ASA), the choice being partly determined by the type of colour film used for the rest of the programme. Colour negative stock offers only one type, which is rather fast (100 ASA), but being a negative stock there is a great range of colour balance in the printing stage.

To determine the calibrations for pans and zooms, it is necessary to put the artwork in the start position, and note the counter numbers. Now move it to the finish position and again note the counter numbers. Subtract the higher from the lower, to find the unit difference between the two positions. Now work out the number of frames to be shot for this action, and then divide the unit difference. The resulting figure will give the counter calibration for each frame.

EXAMPLE

counter reading in start position	=	0000
counter reading in finish position		0240
unit difference		240
number of frames to be shot		24 frames
counter calibration 240/24	=	10

Usually, counter calibrations do not work out to a complete number and a fraction is involved. The normal procedure is to take it to the next number ($10\frac{1}{2}$ becomes 11, for example), which means that the total calibrations would be more than required. To counteract this, the first few counter calibrations and the last four are halved, which normally gives the correct number of calibrations. A further reason for doing this is that it starts and finishes a movement more smoothly, thus preventing a visual jerk. When working out movements be careful that the counter calibrations are not too large, otherwise there will be jerky movements and the image will probably move across the screen too fast. Generally when counter calibrations are large, it means that the proposed movement (the distance between the start and finish position) is too long in relation to the number of frames to be shot.

If you make a mistake while filming, close the camera shutter and shoot a dozen frames, which will come out black on the print and indicate a new start. Then start filming the sequence or shot again, starting it from a cut between shots. A new start cannot be made with a mix, so it may mean refilming more than the sequence with the mistake. Always add at least six frames to the first shot to allow for frames lost when editing, and the same on the last shot. Also shoot a dozen frames black before doing any more filming. If you ever have to stop filming in the middle of a sequence, make a note of exactly how far you have got, and note all the counter readings and the light settings and exposure. This is very important, just in case anyone touches the controls or artwork and changes their positions. Your memory is not sufficient in situations like this.

If your camera does not have a variable shutter, then fades, mixes and superimpositions will have to be achieved in the negative cutting stage by cutting the master into A and B rolls. However, to accommodate fades and mixes, extra film must be shot to allow for the length of the fade or mix. If the mix between two shots is 24 frames, then both shots will need 30 additional frames when filming.

Exposure reading procedure can vary according to the type of artwork. If the artwork is white line or letters on a black ground, take an exposure reading from a sheet of white paper; this will give the minimum exposure for the white and the black will be solid. If the artwork is mainly black lines on a light ground, take the exposure reading from a piece of light grey card. This will prevent over-exposure of the black lines. When photos are used, take a reading from each print, just in case there needs to be an exposure change. Never guess; the eye is easily fooled, and this could ruin a sequence. The shutter speeds of cameras vary; special animation cameras are normally set at $^{1}/_{10}$ second, whereas cameras which can film both live action and animation are set at $^{1}/_{30}$ second for animation. Some cameras have coupled light-meters, and when filming animation the control needs to be set so that the film's ASA speed corresponds with a camera running speed of 16 frames per second. This will give the correct exposure.

LIVE ACTION TITLES

Some title sequences are designed as live action film, rather than captions or animation. These are often used for plays, where the sequence is usually a lead-in to the first scene and presents some of the characters. In other cases it is a self-contained unit which introduces the lead characters, or shows an accumulation of actions associated with the hero or subject. In many cases the titling sequence is the first scene of a play, with lettering superimposed at appropriate points, but obviously the sequence will not have been the designer's creation; it was produced by the director of the programme. This may be because the director wants the titles to be an integrated part of the whole programme or to create a particular mood.

Live action titling is considerably more expensive than any other form of titling. Either the film has to be shot specially, which may mean the expense of transporting a camera crew and actors to a location, or the sequence must be made up of library film. Apart from news or historical events, most library film is not exactly what is required; the camera angle may be wrong or the shot too wide. In order to design a sequence the designer must take into account that the exact shots may not be available and he must have sufficient flexibility to modify his idea.

All the criteria which apply to other forms of titling sequences also apply to live action titles. In particular, the clarity and simplicity of the shots is very important because the inherent movement in the shots can confuse the viewer. Perfecting a movement or action can involve many 'takes' and varieties of shots, so that there is some choice at the editing stage.

Some of the most successful sequences have controlled movement within the frame, or the movement is seen in close-up, thus concentrating the viewer's attention. To accentuate the movement the freeze-frame technique, in which a significant action is frozen on the screen for a few seconds before it continues normally, is used. Another

Five stills selected from a live action titling sequence, showing the necessity of using simple camera movements and focusing, clarity of letter form and sensitivity in the positioning of the information on the film.

method is to use slow or fast motion, or even time-lapse, where film is exposed a frame at a time at intervals ranging from one minute to twenty-four hours, thus telescoping time dramatically.

Added to these techniques is the potential of split-screen images, where two or more live action shots occupy a different portion of the screen at once. This form of presentation is more complicated, not only because it needs careful planning if the image on the screen is not to become confusing and unclear, but also, and more importantly, because it requires expensive optical printing.

Whatever the final form of the sequence, the designer must be very clear about what he wants, because a film cameraman must be instructed to shoot the film and record sound, and directions must be given to the film editor. Neither of these technicians can wait for you to decide on a shot, because time costs money and must be budgeted for.

Live action sequences can be conceived with synchronized sound which can be recorded while filming, whether dialogue, actuality sound or sound effects. Good sound is important for titling sequences; it helps to set the mood and music can be used to establish a rhythm. The devising, shooting and editing of live action title sequences requires considerable skill and co-ordination, so it is an advantage if the designer has had some film-making experience. In some small television organizations it might be necessary for him to do the filming and editing as well as the designing.

This brings us back to an earlier point: the wider the designer's experience, the more ideas, approaches, methods and thought he can bring to bear when designing for television. The range and scope of the work are so wide, the demands are so great, the medium consumes so much material each week, that the designer must always be looking for new ideas, particularly in programme titling. Programme titling is one of the most demanding aspects of the designer's work, and because it involves time, movement, sequence and sound, it is much more difficult to master and very difficult to describe with still pictures and words. Good design ideas in this area come by practice and experience; the solution of yesterday's problems provides an unconscious basis for solving the problems of today.

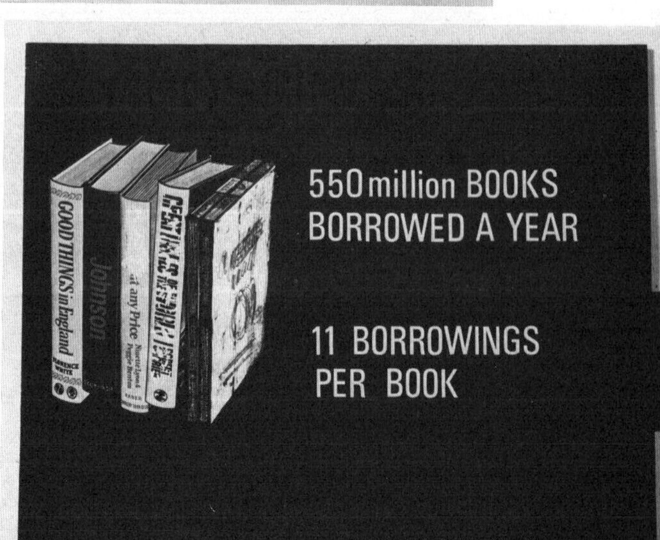

These three illustrations give some indication of how a television animation works. The caption starts with some information and then further information is revealed while the caption is still on air. The hidden letters are covered by a piece of black card, called a 'slide', and this is pulled or pushed to reveal the letters.

6 Television animations

Television animations are single graphics which have an element of movement introduced while they are on the screen. The movement is built into the caption and its artwork, although further movement can be achieved by superimposing a related second caption over the first. There are several ways of producing such animation, all making use of the inherent qualities of electronic pictures and having nothing to do with film. Similar results are obtainable with film animation, but television animations can be produced almost instantly. Film requires more complicated artwork, filming, processing, etc., whereas television animations require only artwork and can be transmitted within seconds; all that is needed is a rehearsal for the director, cameraman and presenter.

TV animations rely on the electronic device of 'black crush' – the altering of the black end of the tone scale by electronic means. This enables several layers of black card to be used in a caption; the edges and different thicknesses cannot be seen on the screen as they are solid black. The image (lettering, drawings, diagrams) should be white, as this is not affected by the black crush, and unless the white is rather transparent, opaque white should always be used. The process of television animation is, simply, that some information on a caption is initially covered by a piece of black card, which, at a given signal, is removed to reveal the hidden information. The result on the screen is either a fast or slow growth of information which is added to the image already on the screen, as though by magic.

Animated captions can be valuable in a programme because they are a quick and efficient way of building up information on the screen, of giving a step-by-step explanation, of increasing the impact of an image, and of emphasizing a particular point, as well as perhaps providing an element of comparison or surprise. They allow the addition of images and information without making another caption, and therefore do not require a change of shot by cutting to another camera source.

They are used considerably in news and current affairs presentations, children's, general interest and educational programmes, and in titling for a wide range of programmes. They are generally produced quite quickly and, when used properly, are an important element of these types of programme. Although the principle of animated captions is

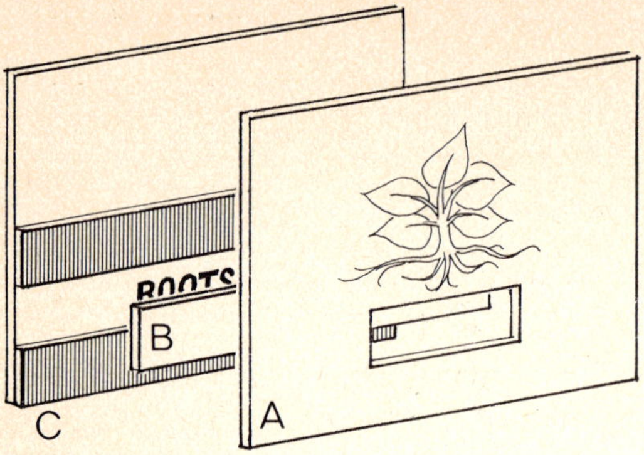

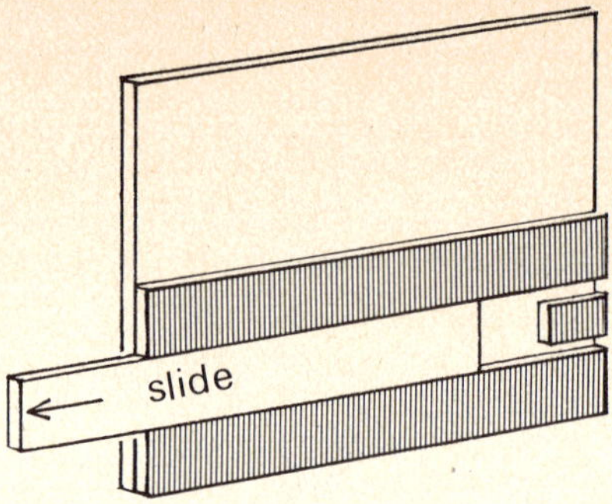

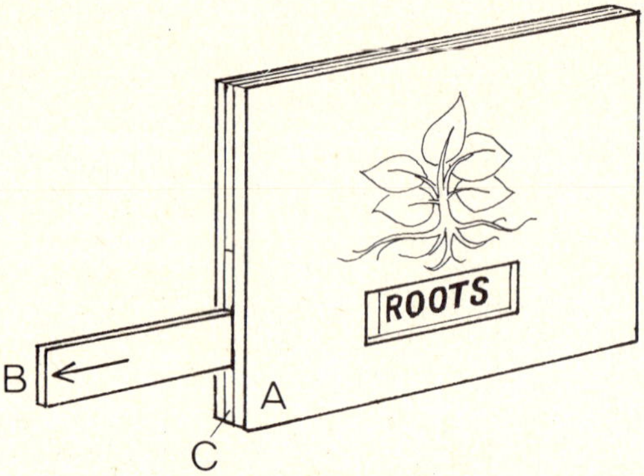

The diagrams above show the basic construction of a television animation. It is made from three pieces of black card which are sandwiched together. Card A has the initial information to be seen on the screen drawn in white or light grey plus a window cut-out; card B is used to make a slide and two runners; and card C contains the information to be revealed, also in white. All TV animations which are white information on black card rely on the principle of 'black crush', an electronic facility which makes joins and edges on black disappear, and the information stay white.

very simple, their production is a little complicated because the movements must be accurate and confined to a 12 in. × 9 in. size, thus fitting into a caption stand. Generally they are not operated by the designer, but by the studio floor manager or a charge-hand, so it is necessary to conform to certain standards. They must not fall apart during operation.

The basis of the animated caption is three levels of black card which are sandwiched together. The top card A contains some information (the initial image on the screen) and also some holes or windows cut out at specific places. The middle card B is cut into a strip which is slightly wider than the vertical height of the window and is called a slide. This is held in place by runners, or strips also cut from card B. The bottom card C contains additional information which is positioned in relation to the window in card A. The slide should be laid on card C so that the information is covered up, and its position marked. The runners are now stuck to card C with double-sided Sellotape. Do not press them too hard against the slide, which otherwise would be difficult to pull. Now lay card A on top, to check that the slide is in the correct position in relation to the window. Card A is fixed to the rest by applying double-sided Sellotape to the top of the runners and staples at the corners. The finished animation will be 12 in. × 9 in. with a piece of card projecting out from one side, which is the slide.

That is the basic mechanism of slide animation, but there are several other factors which must be observed in its production. Firstly, it is important to cut the edges of the slide parallel, otherwise it will jam when pulled. It is also useful to bevel the edge of the slide, so that it does not scrape

The top set of diagrams shows a way of animating a walking figure across a background. In this case the background stays static and the figure walks across the screen. The figure is composed of one cel with the head and body. There are a series of cels (probably twelve) which cope with the movements of the arms and legs, and are laid on top of the body. These two are registered on a peg-bar which is geared to move, in this case, from left to right. The filming entails changing the legs/arms cel every one or two frames, and winding the moving peg-bar every frame.
The second method is slightly different, because the background moves and is registered on the moving peg-bar, and the cels for the figure are on static pegs. The background is moved from right to left to create the illusion of the figure walking.

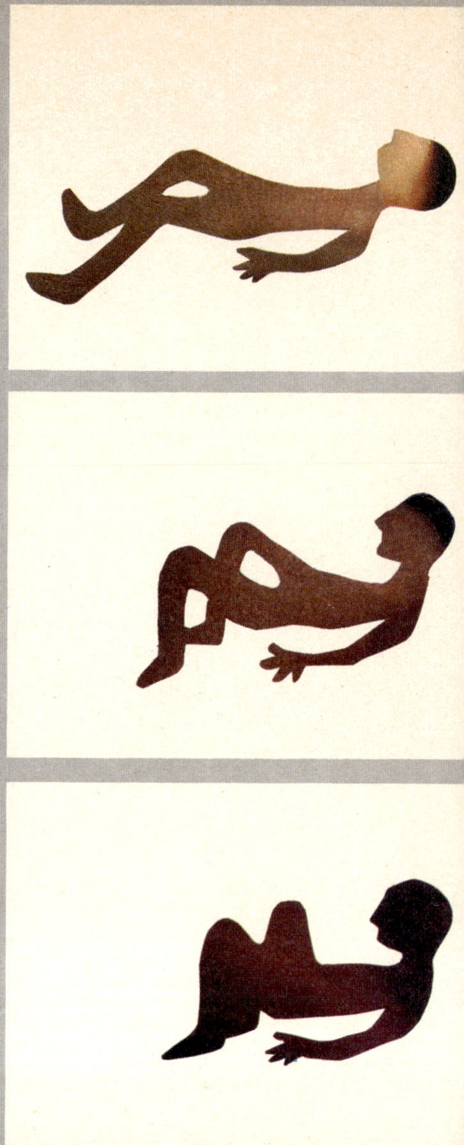

These are key drawings from animated sequences, using a fluid, stylized form of drawing. There are at least four in-between drawings filmed between these keys, providing a smooth metamorphosis.

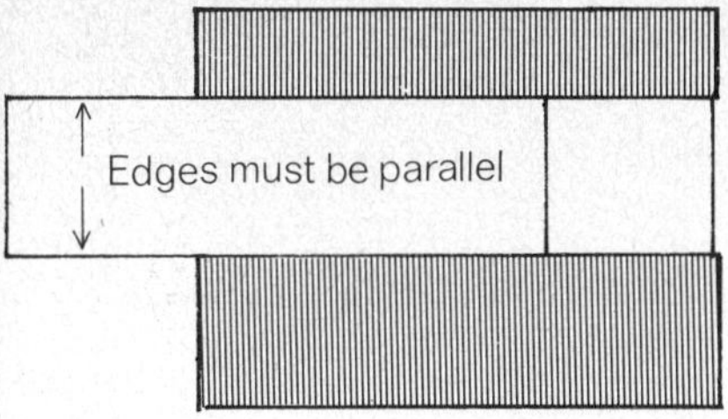

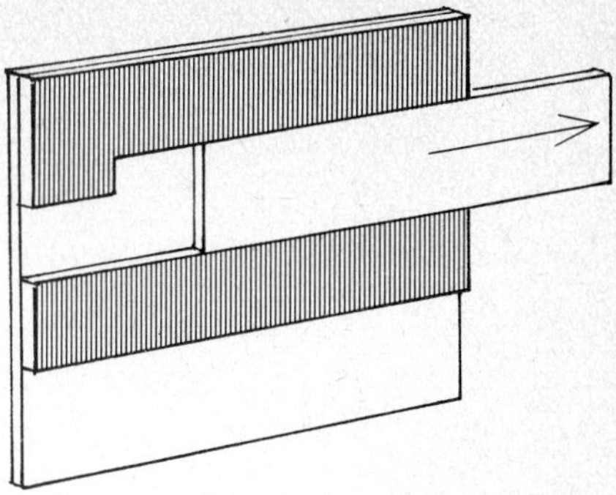

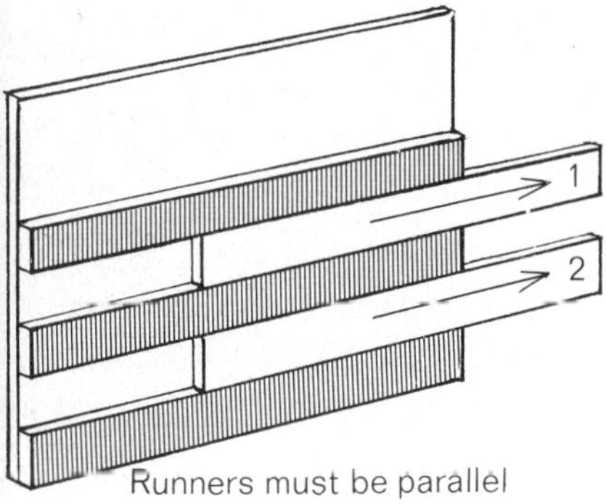

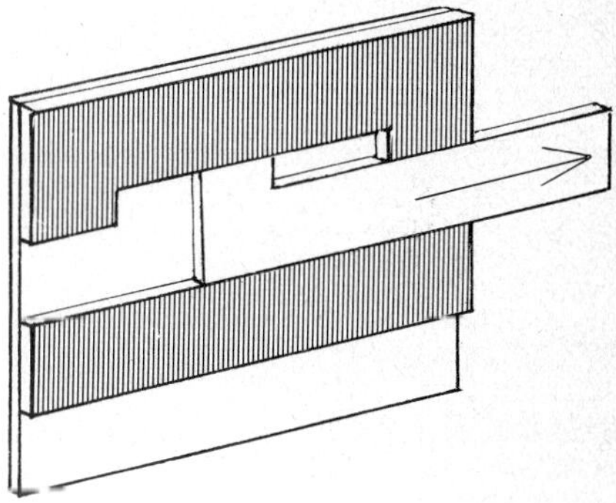

the artwork on card C. Secondly, the edges of the window cut in card A, and any edges visible on the slide as well, must be blacked with a fibre-tip pen. Thirdly, it is important to make sure, during both the preparation of the artwork and the construction of the animation, that the information on card A is parallel to that on card C, particularly if it is lettering. The construction of the animation must be clearly thought out before you do the artwork, because it is easy to make a mistake in cutting the separate pieces, perhaps then finding that they will not fit together or that something is misaligned. Another tip is to fit a stop on card C, so that the slide does not disappear inside; the stop is fixed in the same way as the runners. An alternative method is to cut an indentation in the runner and have an extra width of card at one end of the slide. This stops the slide at both ends of its travel but requires more accuracy in cutting the slide and runners.

It is possible to have several slides in one animation; the number will be limited by the amount of information that needs to be displayed. When lines of lettering are animated, the limit is four lines to be revealed by separate slides on one 12 in. × 9 in. caption, giving a total of five lines on the screen – an animation should always start with at least one line of lettering on the screen. Lines of lettering should always be revealed from left to right, the normal scanning direction when reading. When this is not possible, then

(*Top left*) To ensure that the slide moves smoothly, its edges must be cut parallel, and the runners must be straight.
(*Top right*) The long notch cut in the top runner prevents the slide from being pushed too far in one direction.
(*Above right*) In this case there is an indentation in the top runner, and an extra width of card at one end of the slide. This limits the travel of the slide in both directions.
(*Above left*) When more than one slide is required, not only the edges of the slide have to be parallel, but the edges of the runners as well. There is a common runner between the two slides.

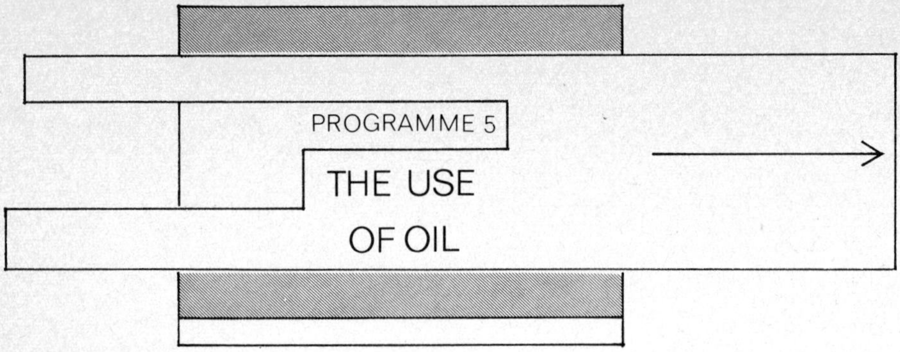

By using a very long slide cut (*above*) it is possible to reveal several lines of lettering one after the other, rather than having two or three separate slides.

they must be revealed by a vertical slide pulled either up or down. Where more than one slide is used, it is possible to use a common runner between each slide, but they must be cut parallel so that the slides do not stick. Always remember to number the slides so that the operator pulls them in the correct order, and to put an arrow to show the direction of pull, because sometimes slides are pushed in and not pulled. When designing animated captions, be careful not to cram too much information into the 12 in. × 9 in. area, because not only will it be difficult to fit all the necessary slides and runners, but the design will probably be rather cluttered. Also, the cut-off area must be considered, otherwise some of the animated information will be lost in cut-off. The simpler the design for an animation, the better it will communicate. Sometimes scriptwriters and producers want too much information put into an animation; the designer should then recommend that it be simplified or made into two animated captions.

Apart from having separate slides, each revealing one line of lettering, it is possible to cut one slide which will reveal several lines of lettering, one line at a time, the only problem being that the slide has to be rather long and cut in steps. This method, however, saves having to cut several slides and runners, which can be time-consuming.

Slides can be horizontal or vertical, or even a combination of both, depending on the information to be presented. A variation on the normal method is to construct the animation as described, except that a window is cut in the slide, slightly larger than the image to be revealed. The window is positioned some distance from the end of the slide, so that initially the image will be covered. To reveal the image the slide is pushed down or in, its travel being limited by a stop. An extension of this, when a series of images need to be revealed and then removed one after another, is to use a slide with a window just larger than each individual image. The slide needs to be much longer than usual, as it is operated by pulling or pushing in a series of movements, stopping at each new image until complete. This slide can be horizontal or vertical.

Another method is to have lettering or images on the slide so that when it is pulled it reveals some new information. This is particularly useful when there is an ani-

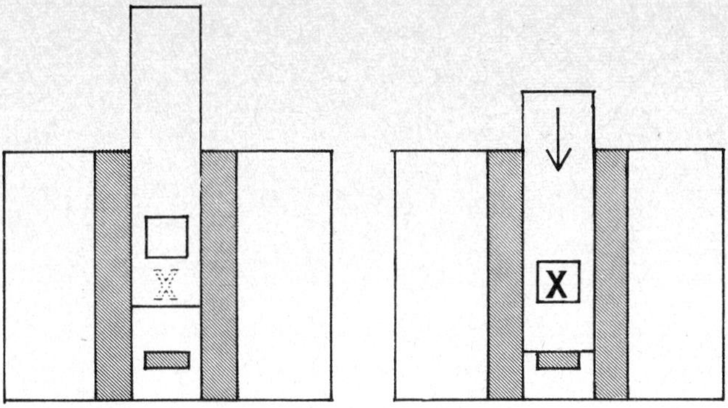

Cutting a window in the slide (*left*) makes it possible to reveal words quickly, by pushing the slide down. There must be a stop to limit its travel, otherwise the word will be covered again if the slide is pushed too far.

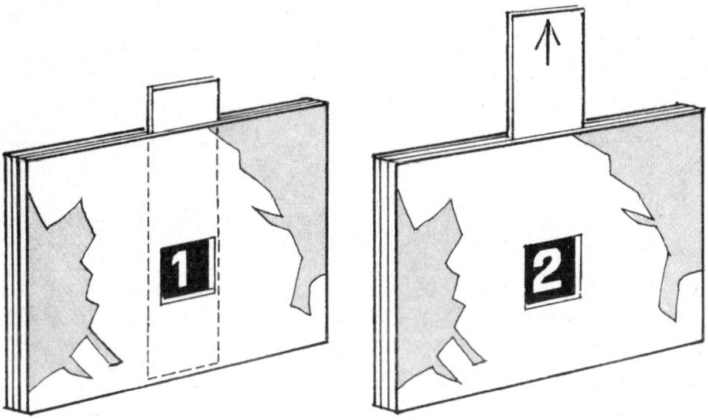

(*Left*) When information needs to be animated on the grey land mass of a map (it cannot normally be done because edges and joins show on grey) a window is cut out of the top card. The slide is black with white letters on it. When the slide is pulled it reveals new information on the card underneath, also white on black.

(*Above left*) Animations in the black sea are basically the same as normal animations.
(*Above right*) Sometimes it is necessary to reverse the conventions for maps, and in these cases there is no problem in having an animation on the land mass.

mation on a diagram or map which has a toned background. Normally it is not possible to have an animation on a toned background, because the different levels would show as shadows which can only be partially removed by electronic means. However, if there is a black rectangle or shape on the toned background, then the window can be within this black area, and a limited amount of black crush can be applied to remove the edges. If a long slide is used, it could have several sets of information on it, which are revealed one at a time. If maps have areas of black background (sea or rivers), then it is possible to have an animation, so long as it is contained in the black area. If animations are required on land masses of maps, then it is necessary to reverse the

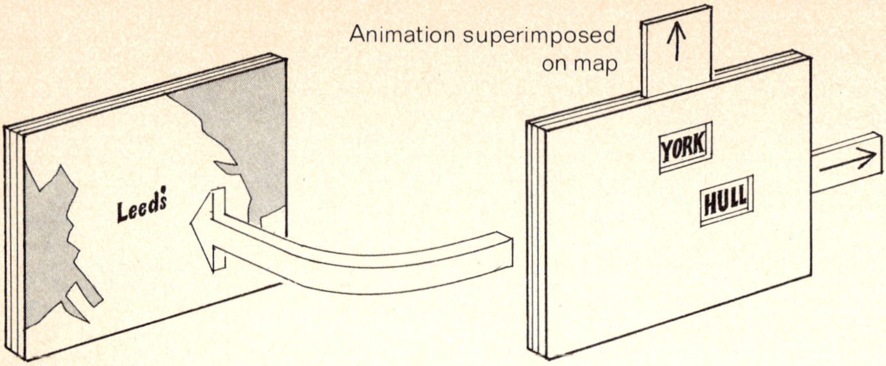

Another way of animating words on the grey land masses is to have a separate white on black animated caption, with the information very accurately matched to the map. The animated caption is superimposed on the map and the words animated on cue. This method is extremely good, but it means that two cameras are needed.

normal procedure, and have a black land mass and grey background sea area. The new information revealed can be white on black (which unifies with the black land mass), or black on white background. If the latter method is used, then the window must be cut out accurately and square with the bottom of the caption, because it will appear as a white rectangle or shape with black lettering or images, and will contrast strongly with the black land mass.

Instead of having a black land mass, the map (or diagram) can be produced as usual, with the animated information – white on black background – on a separate caption which is superimposed or inlaid on the map. This means that white information can be added to the grey or toned areas of the map very successfully, but the animated caption must be very accurately matched when producing the artwork. The only problem is that this method ties up two cameras and so can present some production problems. This method is standard procedure for colour captions, where there is no black, only colours (or tones when reproduced on black-and-white receivers), and where, therefore, black on white animation cannot be used within the map caption.

This method can be adapted for animating information on still photographs or even film and live action pictures. Matching the animated image to still photographs is no problem, but it is considerably more difficult with film and live action. With film it is possible to obtain a few frames which can be placed in an enlarger, and projected to 12 in. × 9 in. size, thus getting accurate matching. With live action pictures, the designer has to assume the positioning of the image (unless given directions by the producer), and the cameramen and director must try to match the respective images by pre-viewing just prior to transmission. To help this matching, the cameraman can have another camera's image electronically fed to his viewfinder, so that it is superimposed on the image of the animation being taken by his camera. This is called 'mixed feed'. If the camera has a zoom lens, it is possible to match the images in a few seconds.

Apart from having animations which reveal information in straight lines, it is possible to reveal images diagonally, either by cutting the end of the slide at an angle (but pulling

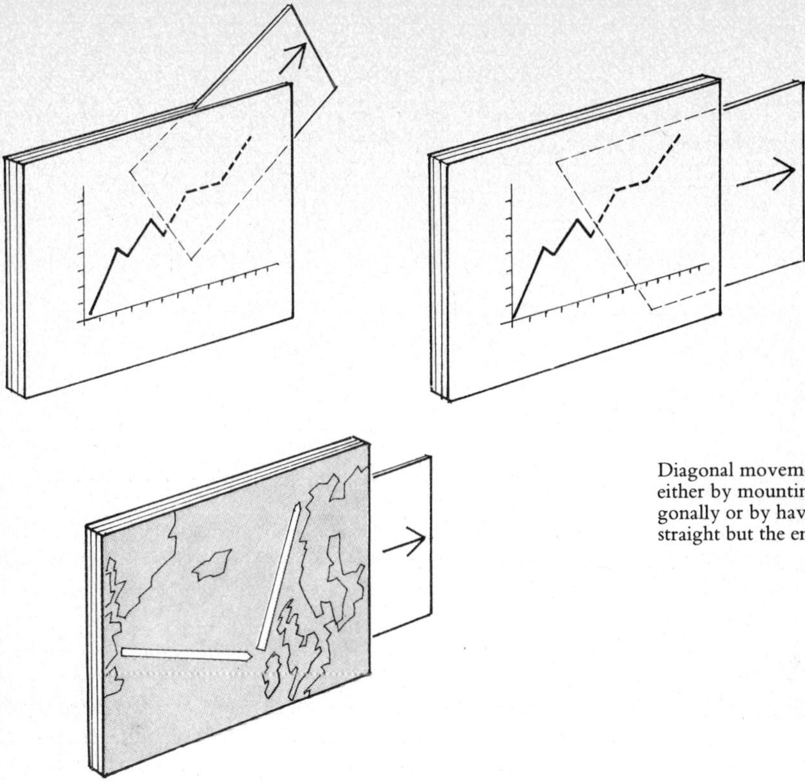

Diagonal movements are possible, either by mounting the slide diagonally or by having the slide straight but the end cut at an angle.

it horizontally or vertically), or by mounting the slide at an angle. The problem of the latter method is that it protrudes at the corner of the caption and therefore will not fit into a normal caption stand. The diagonal reveal is useful for maps showing routes or zones and for diagrams such as graphs.

So far all the animations illustrated have had rectangular windows on the top card, but it is possible and sometimes desirable to use other shapes, particularly when the animation is complex. For example, if an object has to change shape, say from a circle to a square, the square is cut out of the top card, and the circle is drawn on the slide in white. The bottom card is white, and when the slide is pulled quickly, the circle becomes a square. Another example is where three arrows need to change direction. In the starting position, three

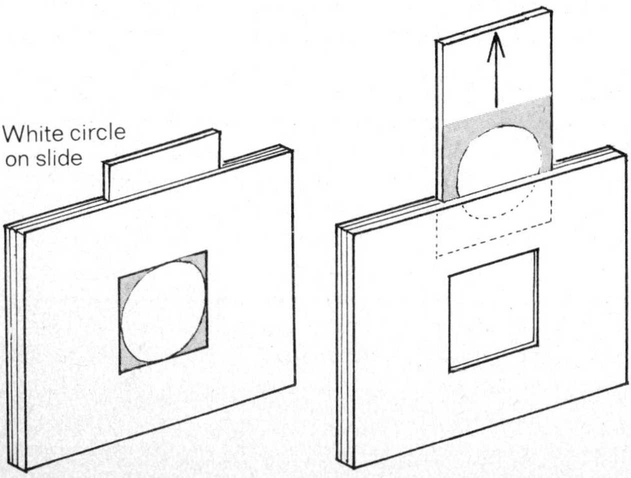

With a white circle painted on the slide, and a square window cut in the top card, it is possible, by pulling the slide quickly, to change the circle to a white square.

This animated caption has both drawing and window cut out of the slide in order to achieve the change in direction and size of the arrows.

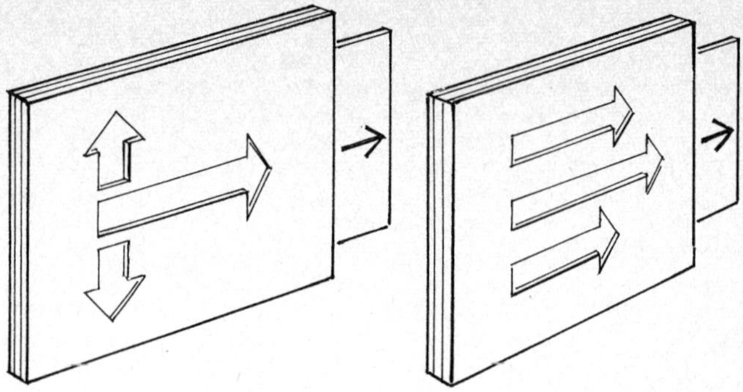

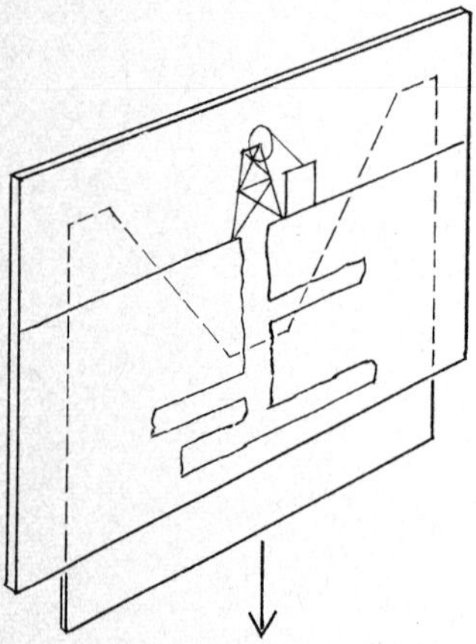

By having a V cut into the slide, it is possible to have the mine shaft revealed downwards and the galleries revealed horizontally.

arrow shapes are cut out of the top card, plus some accurately drawn white areas on the slide. To accommodate the finishing position, the middle arrow window is extended, while arrow shapes are cut for the upper and lower arrows. The new arrow positions are blacked on the slide in its start position and then the white is extended on the slide for the middle arrow's finishing position. With stops fitted, the slide has only limited travel. By pulling the slide quickly, the arrows change dramatically. This could have been produced as two captions, but the problem in hand demanded only one caption.

Another example makes use of an irregular-shaped window cut in the top card, which shows how a mine is sunk and the galleries created at several levels as work progresses. In order to show the correct development of the galleries branching from the central shaft, the slide has a large 'V' cut into it, so that it reveals the shaft first and the galleries grow out of the shaft as the slide is pulled down. The galleries could have been revealed by separate slides, but the operator would then have needed two pairs of hands to co-ordinate the movements. This method can be applied to a variety of subjects.

Another way of revealing or removing small images or words is to use a flap which is stuck on to the caption and can be either lifted or dropped while the caption is on air. This is a very quick and easy method, but it can only be used when the caption design is very simple; it takes practice to lower the flap quickly without shaking the whole caption.

Instead of using slides and windows, information can be hidden by a flap, hinged to the card. The flap is lifted up by the protruding corner (so that the operator's fingers do not get into the shot), to reveal the word when required.

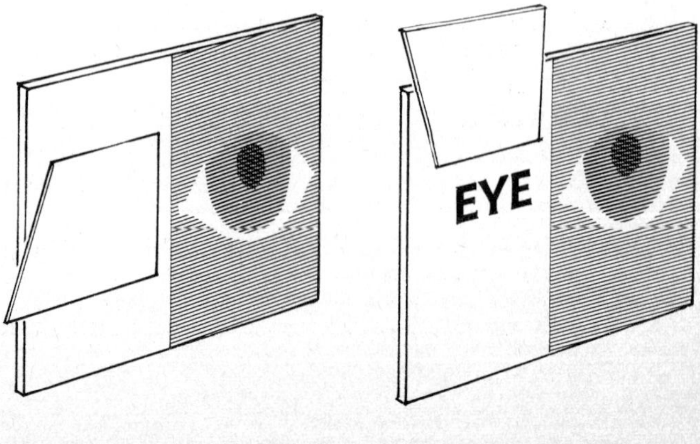

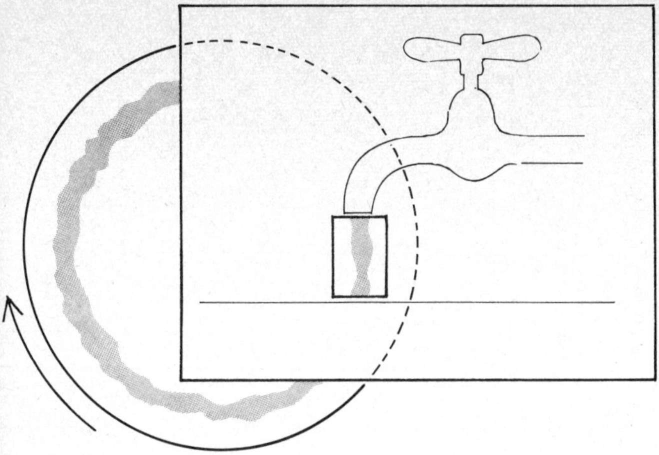

If an animation needs to keep going for more than a few seconds, a circular disc attached with an eyelet to the back of the caption can be the answer. The information is seen through the small window and the operator just keeps turning the disc. In examples such as this it is very effective.

Sometimes an animation requires a movement which must go on for more than a few seconds. If the movement is fairly simple this can be achieved by using a circular disc, with an image that is seen through a small window. It can be turned continuously for minutes if necessary. The disc is attached by an eyelet or by a split paper-clip, and is normally fastened to the bottom piece of caption card, so that the clip is hidden. If it is impossible to hide the clip, then stick a small piece of black card on top, or some matt black sticky tape.

With slide animations, the pushing or pulling of the slides must be positive and deliberate, as jerky movements destroy the illusion. The secret is to be relaxed. Do not snatch the slides; pull them exactly horizontally or vertically, so that they do not jam. If a slide movement has to be fast, then design the animation so that the slide is pushed down vertically. In the case of a line of lettering, it is much quicker to reveal the height of the type than to pull the slide sideways to reveal it letter by letter. It is always safer to tape the animated caption in the caption stand with a few strips of matt black tape, just in case the operator moves the caption when pulling the slide. Also if there are two or more slides to be pulled in quick succession there will not be time to hold the caption in place.

Before transmission, always check the animated captions to be sure there are no unwanted white edges which should be retouched with a black fibre-tip pen. During rehearsal, it is advisable to show the cameraman the animation in its finished position, and then reset it. This enables him to line up and frame his shot without fear of some of the information being in the cut-off area.

When designing animated captions, instead of using words for all the information, a carefully chosen drawn or photographic image can be used with great effect, thus creating some variety in the imagery. This is particularly appropriate for news and current affairs presentations as well as educational programmes. It makes more use of the potential of television imagery, and helps to present the information more clearly and quickly. The images should

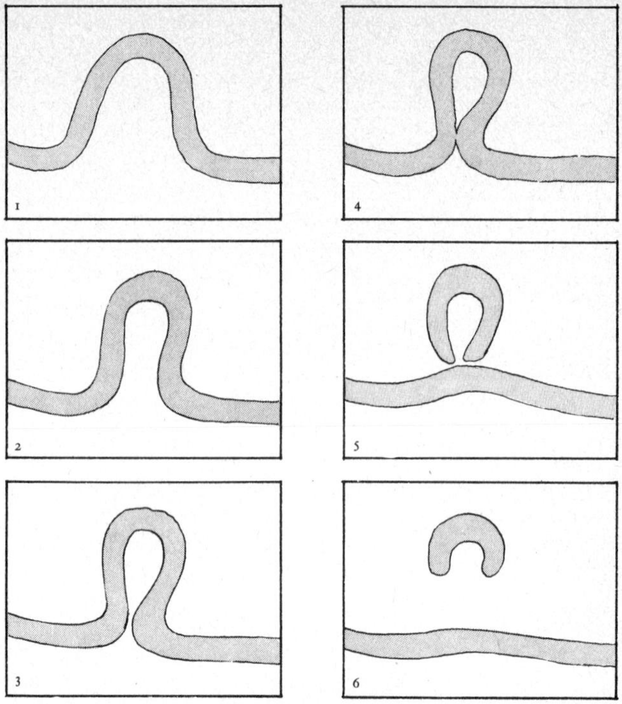

Using a series of captions and slowly mixing from one to the next, a simple form of animation can be produced. The captions need to be split between two TV cameras, nos. 1, 3, 5, etc. for camera one, and nos. 2, 4, 6, etc. for camera two. While one camera is on air, the caption is being changed on the other. The only problem with this form of animation is that the drawings and the camera shot must be lined up very accurately.

be chosen carefully to ensure they are bold and not too complex.

Another form of television animation involves the use of a very carefully matched series of images on separate captions, arranged on at least two caption stands, and mixing or cutting between two or three camera sources. Depending on the point to be illustrated, the basic image can grow or diminish, or move from one side of the screen to the other. This is a modified form of simple cel animation; the image does not move on the caption, it only appears to move through the juxtaposition of shots. Quite complex animation can be devised by this method, although there is a limit to the variety and number of movements that can be sustained at one time. In producing the artwork for these sequences, the images must be very carefully worked out and positioned on the caption to make the movement convincing. It is also an advantage if the cameramen have mixed feed on their viewfinders, to eliminate any variations in framing-up the captions. This method can also be used for still photographs, particularly if they have been shot against a plain background, or cut and remounted on a tone background. This form of animation has a more fluid quality than the slide method; it can also make use of several tones and colours and irregular shapes and forms, which is difficult with slide animations. The images can even be three-dimensional in form and make full use of perspective. This can be a very cheap form of animation because there are no film stock and processing charges, only the cost of the artwork, and it is straightforward from the studio production point of view. The one limitation of note is that each caption has to be changed after being shot, and the floor manager must take it out of the caption stand to reveal the next image. This takes only a few seconds, but it prevents really fast cutting between images.

The production of television animations is an exciting task for the designer, because he is being continually presented with new problems to solve. He is striving to create more sophisticated movements and methods, while trying to cut down the length of time it takes to make an animated caption. Television animation is a means of swiftly transforming static information into images of movement and growth. It can be designed and produced much faster than film animation. It is a vital form of television graphics, contributing to the immediacy and impact of television. Good animated captions can help to liven up the presentation of information on television, and certainly can make that information easier and clearer to understand.

Further reading

Bryne-Daniel, Jack, *Grafilm: an approach to a new medium.* London and New York, 1970.

Callaghan, Barry, *The Thames & Hudson Manual of Film-Making.* London and New York, 1972.

Davis, Desmond, *The Grammar of Television Production.* London, 1967.

Garland, Ken, *Graphics Handbook.* London, 1966.

Halas, John, and Roger Manvell, *The Technique of Film Animation* (2nd rev. ed.). London, 1968.

Laughton, Roy, *TV Graphics.* London and New York, 1966.

Levin, Richard, *Television by Design.* London, 1961.

Lewis, P., *Educational Television Handbook.* New York, 1960.

Madsen, Roy, *Animated Film: concepts, methods, uses.* London and New York, 1969.

Millerson, Gerald, *The Technique of Television Production.* London, 1969.

Spear, J., *Creating Visuals for Television.* Washington, D.C., 1962.

Wright, Andrew, *Designing for Visual Aids.* London and New York, 1970.

Glossary

ANIMATED CAPTION A 12 in. × 9 in. caption with slides and flaps which conceal some information. The slides are pulled or pushed to reveal the information, at a given cue. It is a quick and simple way of obtaining movement on a caption.

ANIMATED TITLE Lettering or other images which have a certain amount of movement on the screen. This can be achieved by film animation, by panning and zooming of a TV camera over captions, or by animated captions.

ASPECT RATIO The ratio of the vertical edge of the frame to the horizontal edge of the frame. The ratio for television is 4:3, and the standard caption size 12 in. × 9 in.

BACK PROJECTION A method of presenting images on a screen which is integrated with the studio set. These can be produced with film or slides.

BACK-RIM LIGHTING This is a means of giving a slight halo effect on a performer or object, by placing a light behind the person and directing it on to the top of the head. This gives more depth to the picture, and separates the person from the background.

BARN DOOR Metal flaps on hinges, which are fixed to the front of spotlights. By adjusting them – covering part of the front glass of the lamp with the doors – the size of the light beam can be adjusted.

BLACK CRUSH An electronic facility which alters the tonal structure of a television image, giving a very contrasty black-and-white picture without any half-tones. This is done by lowering the brightness (lift) of the TV image. This facility is very important when captions are being supered, and also for TV animations.

CALIBRATION A method used in film animation, and also a unit of measurement. Animation rostrums have counters which measure the various movements of the table and camera, in units of 100 to an inch. When movements are required, the operator counts the number of units and divides it by the number of frames for the shot (in other words, calibrates it), which then gives the required number of units, or calibrations, for each frame of film.

CAPTION — This is the term for all forms of artwork, lettering, photographs and diagrams which are produced on card or paper 12 in. × 9 in., placed in a stand in the studio, and shot by the TV camera.

CAPTION SCANNER — A fixed-focus camera for transmitting captions, instead of a studio TV camera. Some scanners have a mechanism for changing the captions.

CEL — Transparent plastic sheets on which images are drawn for film animation.

CONTACT PRINTING — This is a method of printing a photographic image to exactly the same size as the negative. The negative is placed on top of a sheet of photographic paper and held in close contact by a sheet of glass. These are placed on the baseboard of an enlarger, and exposed for the required number of seconds.

CREDIT — An acknowledgment of a performer, or a person who has contributed to the programme, usually done by lettering.

CRUSH — See BLACK CRUSH.

CUT-OFF — The area of a picture which is lost during the various stages of television production, before it is received on the domestic TV set. On a 12 in. × 9 in. caption this represents a margin of almost 1½ in. All vital information must be contained in the centre area of approx. 9 in. × 6 in.

DOPE SHEET — A chart on which are written all the instructions for filming animation. These detail how many seconds or frames a shot should be, and whether pans or zooms, mixes or fades are needed.

DRY-MOUNTING — A method of sticking photographs and drawings to stiff card. A material called dry-mounting tissue is placed between the photo and the card, and then they are put into a dry-mounting press, which applies pressure and heat, bonding the photo to the card.

FIELD AREA — The field area is a chart marked out on cel, which, when laid on top of artwork for filming on an animation rostrum, will show the framing of the camera, and indicate how near the camera needs to be to the artwork.

FOUNDERS AND BRASS TYPE — The metal type which is normally used for printing texts in books and display posters.

HOT-PRESS LETTERING — This is lettering which has been printed by using pressure and heat, the metal type being applied to a pigmented foil, which adheres to the caption card. It is very similar to typographic printing, except that foil is used instead of ink, and the type is heated.

A method in which the picture from a TV camera is inserted into an area of a picture from another TV camera. — INLAY

This is lettering made in various sizes and styles, which by rubbing with a pencil, will adhere to many surfaces. It is sold under brand names such as Letraset, Letter-press, etc. — INSTANT LETTERING

This is a very high-contrast, slow ASA speed negative sheet film, used mainly for copying black-and-white subjects, or making duplicate negatives. It reduces the subject to harsh black and white, with no half-tones. — LINE FILM

A gradual merging of one image into another, till the first image has disappeared and only the second remains. — MIX

A movement of a camera to the left or right, while taking a shot of a caption or view. — PAN

Metal pins used for the accurate registration of artwork and cel, in the drawing and filming stage. They correspond to holes cut in the cel by a special punch. The usual system is one small round hole, with two rectangular holes punched 6 in. either side of it. — PEG-BAR

See SCREEN RASTER. — RASTER

A method of developing a photographic negative, which causes the grain in the emulsion to form large clusters rather than very fine dots. — RETICULATION

An electronic facility which converts negative images to positive, or positive to negative. — REVERSE PHASE

A caption which is 12 in. wide, and several feet long, normally used for lettering of credits. The roll is placed in a machine, so that the lettering moves either up the screen or from right to left across the screen. — ROLLER CAPTION

A special cine-camera which is fitted to a vertical column and points down on to a table which will move in several directions. This movement is controlled by gears and counters. The artwork is placed on the table and held flat by a glass pressure plate. The camera shoots one frame at a time, and can move up and down the column, thus changing the framing of the artwork. — ROSTRUM CAMERA

All television pictures are made up of a series of lines or dots, with 405 (or 625) lines traversed every $^{1}/_{25}$ second by a beam scanning the front face of the tube. The line effect produced on the tube, which creates the picture, is the raster. — SCREEN RASTER

A funnel-shaped attachment to the front of a spotlight, giving a very narrow beam of light. — SNOOT

STOP-FRAME ANIMATION — The creation of moving images by filming a series of static drawn or photographic images, one frame at a time.

STORYBOARD — A chart which gives some indication of the visuals, soundtrack and filming directions of a film sequence.

STROBING — Visually disturbing effect caused by the coincidence of horizontal lines with the raster of the TV screen, or when the camera and the subject are moving too fast in relation to one another.

SUPERS — White-lettered caption on black card, superimposed on another picture source.

TELECINE — A film projector which translates the film image into an electronic signal. It is interlocked with a special TV camera and has a separate control panel.

TELEJECTOR — Similar to a telecine, except that it is a slide projector coupled to a TV camera.

VIDEOTAPE RECORDER — A machine which records and plays back video and audio signals. This is the normal way of recording television pictures and sound.

ZOOM — A lens which gives an infinitely variable focal length. It will go from telephoto to wide-angle by the operation of a lever, without moving the camera. In animation, to obtain a zoom the camera moves nearer to or further from the artwork.

Index